Asian

Latest Revised Edition

2022 Syllabus, NSQF Level-3 Compliant

Q-Bank Series
MECHANIC DIESEL

Industrial Training Institute

Skill India
कौशल भारत - कुशल भारत

कौशलं बलम्

Sector - Automotive

As per NSQF Level-3 Compliant Competency based Curriculum prescribed by D.G.T. Govt of India

For Annual A.I.T.T. Exam.

AF658165

NSQF Level-3 Latest Curriculum

Multiple Choice Questions (with key)

Latest All India Trade Test Papers (with key)

MCQs For All Competitive Exams

G.S. Sethi

COMPUTECH PUBLICATIONS LIMITED
An ISO 9001 : 2008 Certified Company
4591/15, Aggarwal Road (Near Happy School)
Darya Ganj, New Delhi - 110002 (India)
Division - AP Asian Publishers
Buy books online : www.computechpub.in

स्वच्छ भारत
एक कदम स्वच्छता की ओर

Latest Edition (NSQF) : **2024**

Price : ₹ **142.50**

ISBN 978-81-7317-767-5

Published by :

Computech Publications Ltd. (*Division* – **Asian Publishers**)
4591/15, Aggarwal Road (Near Happy School)
Darya Ganj, New Delhi – 110002 (India)
Phone : 011-43080162, 43080163, 43080164 *Fax* : 011-43080165
WhatsApp : +91-9716836703
E-mail : nap@newasianpub.in ; info@computechpublications.in
Buy books online : www.computechpublications.in

Preface to NSQF Compliant Revised Edition

Mechanic Diesel under the Craftsmen Training Scheme is one of the popular trades for which training is imparted through a vast network of Industrial Training Institutes. The competency – based curriculum prescribed by the Directorate General of Training (DGT) comprises basically of domain area and core area. Within the domain area, trade theory forms an important part. While trade practical takes major part of training time, trade theory tops the list among all other subjects in terms of training time allocated. Undoubtedly proper understanding of theoretical aspects of the trade helps build a strong base to meet the challenges a craftsman has to face when later settled on the job.

As per the current trend, trade theory related knowledge and understanding is tested through a series of objective type test items, wherein a candidate is required to choose the correct option within those given (usually four) for each question. A candidate can successfully face this type of test only if he can well understand the question and differentiate between the one right option and the other wrong options.

This Q – Bank relating to the Mechanic Diesel Trade has been prepared keeping this basic issue in mind. There are nearly 1000 multiple choice questions, grouped into appropriate modules and in a pattern similar to the one the candidates are required to face in the All India Trade Test. To further enhance its utility and to help trainees to learn through proper understanding, explanatory notes are given to several questions supplemented by quality figures, definitions, data etc. which should prove highly beneficial to the trainees to prepare for the trade test with confidence. Furthermore, latest A.I.T.T. question papers with key have been included.

Besides the ITI trainees in Mechanic Diesel trade, candidates appearing for various competitive examinations / interviews for jobs in many public, private, state and central government enterprises should find this book of great help. Several enterprises prescribe ITI level as necessary qualification for recruitment. Railway Recruitment Board (RRB), enterprises engaged in manufacture, sale and service of automobiles, are among a few such examples.

Wishing our trainee readers a grand success in A.I.T.T. and job seekers decent post-training placement we remain committed to continuous improvement and would welcome suggestions for further improvement of this book.

— Author

Contents

Module 1

Workshop Safety Practice

MULTIPLE CHOICE QUESTIONS

1. The type of occupational health hazard which involves "toxic" is:
   (a) Physical hazard
   (b) Chemical hazard
   (c) Biological hazard
   (d) Mechanical hazard

2. Which type of personal protective equipment is recommended for use, to handle loads having rough surfaces and jagged edges?
   (a) Paper gloves
   (b) Rubber gloves
   (c) Leather gloves
   (d) Polythene gloves

3. What type of gloves is used to avoid cuts and abrasion during material handling?
   (a) Rubber gloves
   (b) Leather gloves
   (c) Cotton gloves
   (d) Polythene thin gloves

4. Burning wood is classed as which type of fire?
   (a) Class A (b) Class B
   (c) Class C (d) Class D

5. Which of these is a biological hazard?
   (a) Smoking (b) Sickness
   (c) Infection (d) Poor discipline

6. What does the following warning sign indicate?

   (a) Risk of fire
   (b) Laser beam
   (c) Toxic hazard
   (d) Risk of electric shock

7. A person coming in contact with live electric wire should be removed from live supply using a:
   (a) Piece of dry timber
   (b) Piece of wet timber
   (c) Metallic rod
   (d) Scaffold rod

8. Which fire extinguisher is used for flammable liquid fires?
   (a) Halon extinguisher
   (b) Dry powder extinguisher
   (c) CTC extinguisher
   (d) Water extinguisher

9. Which of these is a toxic in an automobile workshop?
   (a) Old bearings
   (b) Paper wrappers
   (c) Used lubricants
   (d) Old washers, bolts and nuts

10. A person faints due to suffocation, the first aid to be given is:
(a) Give him a warm drink
(b) Sprinkle cold water on his face
(c) Place him in a well-ventilated location
(d) Give him artificial respiration

11. An oily floor should be cleaned by:
(a) Cotton waste (b) Putting water
(c) Putting sand (d) Putting saw dust

12. What does the following mandatory sign indicate?

(a) Save water (b) Wash hands
(c) Canteen zone (d) Drinking water

13. Which equipment is used in vehicle service station to lift a vehicle for water wash?
(a) Hoist (b) Crane
(c) Stand (d) Screw jack

14. Which equipment is used to lift and support a vehicle before work underneath it can be safely undertaken?
(a) Stand (b) Sling chain
(c) Lifting crane (d) Hydraulic jack

15. Which oil is harmful to the environment?
(a) Engine oil (b) Hydraulic oil
(c) Seeds / nuts oil (d) Used engine oil

16. What can cause fire while refuelling hot engine?
(a) Vapours (b) Paper
(c) Cloth (d) Leather

17. Which gas is harmful to human health?
(a) Nitrogen
(b) Water vapour
(c) Carbon dioxide
(d) Carbon monoxide

18. What is the effect of air borne dust in workshop?
(a) Diarrhoea
(b) Dehydration
(c) Throat infection
(d) Rise in blood pressure

19. Which type of fuel contains more carbon monoxide?
(a) Solid fuel (b) Burnt fuel
(c) Liquid fuel (d) Un-burnt fuel

20. Which automobile component produces dust?
(a) Axle (b) Piston
(c) Gearbox (d) Brake shoe

21. Which material can cause lung cancer?
(a) Fibre (b) Wood
(c) Graphite (d) Asbestos

22. What is ABC in first aid?
(a) Army, Branch, Calculate
(b) Aviation, Breathing, Cumin
(c) Away, Breathing, Calculation
(d) Airway, Breathing, Circulation

23. Which part of body bleeding profusely is considered serious and need professional attention?
(a) Leg of the human body
(b) Knee of the human body
(c) Wrist of the human body
(d) Thigh of the human body

24. How to treat burns and scalds?
(a) Blow hot air on the injured part
(b) Blow cool air on the injured part
(c) Cover the injured part with cool water
(d) Cover the injured part with hot water

25. Which device is best for use to control toxic waste?
(a) Wire brush (b) Cotton waste
(c) Air compressor (d) Vacuum cleaner

26. What comes under first aid?
(a) Treating a victim for a shock
(b) Completing the primary source
(c) Assessing a victim's vital signs
(d) Immediate care and support to the injured person

27. What is your immediate action on completion of first aid?

(a) Call a taxi
(b) Call victim's friend
(c) Call fire service
(d) Call emergency service

28. What is the benefit of 5 S system?
(a) Increase in manpower
(b) Increase in economy
(c) Increase in productivity
(d) Reduction in pollution

29. What is biological hazard?
(a) Virus (b) Fatigue
(c) Alcoholism (d) Old age

30. What is the first action if a colleague (victim) received a heavy electric shock?
(a) Pull the victim from the live conductor
(b) Pour water to avoid burning of victim
(c) Fetch the first aid box
(d) Switch off the current immediately

31. What does 'horseplay' refer to?
(a) Animals (b) Game of horses
(c) Fooling around (d) Throwing hay

32. What three elements must be present to cause a fire?
(a) Fuel + Heat + Oxygen
(b) Oxygen + Fuel + Paper
(c) Heat + Water + Oxygen
(d) Smoke + Fuel + Oxygen

33. Isolating fire from supply of oxygen by blanketing with foam and sand is called:
(a) Cooling (b) Starting
(c) Misfiring (d) Smothering

34. What should not be used on burning liquids?
(a) CO_2 (b) Foam
(c) Sand (d) Water

35. Which type of fire extinguisher is not suitable for electric fire?
(a) CO_2 (b) CTC
(c) Foam (d) Dry powder

36. What is used to extinguish class 'B' fire?
(a) CO_2 (b) Argon
(c) Oxygen (d) Nitrogen

37. Which type of fire extinguisher is suitable for extinguishing class 'A' fire?
(a) Halon (b) Dry powder
(c) Foam or water (d) Carbon dioxide

38. Which fire extinguisher is suitable for extinguishing class 'C' fire?
(a) Foam filled extinguisher
(b) Water filled extinguisher
(c) Dry powder fire extinguisher
(d) Carbon dioxide fire extinguisher

39. Stating "Don't spill the fuel on workplace" can be said to be related to:
(a) General safety
(b) Personal safety
(c) Machine safety
(d) General and machine safety

40. What first aid is required in case of a severely bleeding open wound?
(a) Allow wound to bleed
(b) Lower the wound area below the body
(c) Apply pressure near wound to stop bleeding
(d) Pressure on the wound itself to stop bleeding and avoid infection

41. What will you do if an electric shock victim is unable to release his grip from the conductor?
(a) Make sure the power is turned off
(b) Cover all burns with a dry loose dressing
(c) Place the victim on one side with head down
(d) Ask a bystander to help you to move the victim

42. How is waste oil disposed off?
(a) Hand over back to the customer
(b) Throw it in the drain
(c) Keep it in small containers in remote corners
(d) Collect it in waste oil container and deliver to registered vendor

43. Which type of energy conservation comes under the replacement of old machineries?
(a) Minor energy conservation opportunities
(b) Major energy conservation opportunities

(c) Medium energy conservation opportunities
(d) Very minor energy conservation opportunities

44. Which type of energy conservation opportunity involves stopping of water leakage points?
(a) Minor energy conservation opportunities
(b) Major energy conservation opportunities
(c) Medium energy conservation opportunities
(d) Extra major energy conservation opportunities

45. Which is medium energy conservation opportunity?
(a) Housekeeping
(b) Stopping of water leakage
(c) Renovation of old building
(d) Replacement of existing household appliances

46. In case of an accident, the victim should immediately be:
(a) Asked to take rest
(b) Enquired about the accident
(c) Attended to immediately
(d) Left to himself without treatment

47. What should never be placed over a burn?
(a) Aloe
(b) Water
(c) Cocoa butter
(d) Dry sterile dressing

48. What does the cautionary sign given below indicate?

(a) Guarded level crossing
(b) Unguarded level crossing
(c) School zone
(d) Pedestrian crossing

49. What is the colour code of bin for disposal of plastic waste?
(a) Red (b) Blue
(c) Black (d) Yellow

50. What warning sign is shown below?

(a) Explosion (b) Overhead load
(c) Sudden drop (d) Noise

51. Which fire extinguisher is filled with carbon tetrachloride and bromochlorodifluoro methane (BCF)?
(a) Carbon dioxide extinguisher
(b) Halon extinguisher
(c) Foam extinguisher
(d) Dry powder extinguisher

52. What is the colour code of bins for waste paper segregation?
(a) Red (b) Blue
(c) Black (d) Green

53. With reference to using a fire extinguisher, what does 'PASS" stand for?
(a) Pull, Aim, Squeeze, Sweep
(b) Press, Aim, Start, Sweep
(c) Pull, Air, Squeeze, Spray
(d) Pull, Aim, Squeeze, Spray

54. There are usually certain colours used on safety signboards. Which one is for prohibition?
(a) Red (b) Yellow
(c) Green (d) Purple

ANSWERS

1. (b)	**2.** (c)	**3.** (b)	**4.** (a)	**5.** (c)	**6.** (d)	**7.** (a)	**8.** (b)	**9.** (c)	**10.** (c)
11. (d)	**12.** (b)	**13.** (a)	**14.** (a)	**15.** (d)	**16.** (a)	**17.** (d)	**18.** (c)	**19.** (d)	**20.** (d)
21. (d)	**22.** (d)	**23.** (c)	**24.** (c)	**25.** (d)	**26.** (d)	**27.** (d)	**28.** (c)	**29.** (a)	**30.** (d)
31. (c)	**32.** (a)	**33.** (d)	**34.** (d)	**35.** (c)	**36.** (a)	**37.** (c)	**38.** (c)	**39.** (a)	**40.** (c)
41. (a)	**42.** (d)	**43.** (b)	**44.** (a)	**45.** (d)	**46.** (c)	**47.** (c)	**48.** (b)	**49.** (d)	**50.** (b)
51. (b)	**52.** (b)	**53.** (a)	**54.** (a)						

EXPLANATORY NOTES

1: Toxic means containing or being poisonous material.

5: **Infection:** The invasion and multiplication of microorganisms such as bacteria, viruses, and parasites that are not normally present within the body.

20: Brake dust is caused by friction between the iron brake rotor grinding on the brake pads when a vehicle slows down. This brake dust is then worn away and becomes airborne.

21: When a person inhales asbestos, microscopic fibres can become lodged in his lung tissue. Over many years, they can cause enough genetic and cellular damage to cause lung cells to turn cancerous.

22: The ABC's of first aid are the primary things that need to be checked when you approach the victim, Airway, Breathing, and Circulation. Prior to CPR, ensure that the airway is clear, check to see if the patient is breathing, and check for circulation (pulse or observation of colour and temperature of hands/fingers).

33: There are three principles of fire extinction, i.e. cooling, smothering and starvation as shown in the figure given below.

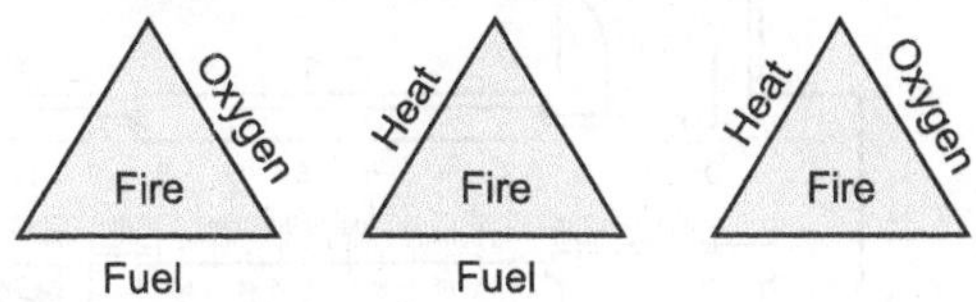

Principle of fire extinction

53: How to use fire extinguisher – use PASS

- Pull the pin on the extinguisher.
- Aim the hose nozzle low toward the base of the fire.
- Squeeze the handle to release the extinguishing agent.
- Sweep the nozzle from side to side at the base of the flames until extinguished

54: The colours used on safety signboards are:

- Red for prohibition
- Yellow for caution
- Green for positive action

Module 2

Measuring and Marking Practice

MULTIPLE CHOICE QUESTIONS

1. What is the name of the part marked as 'X' in the figure given below?

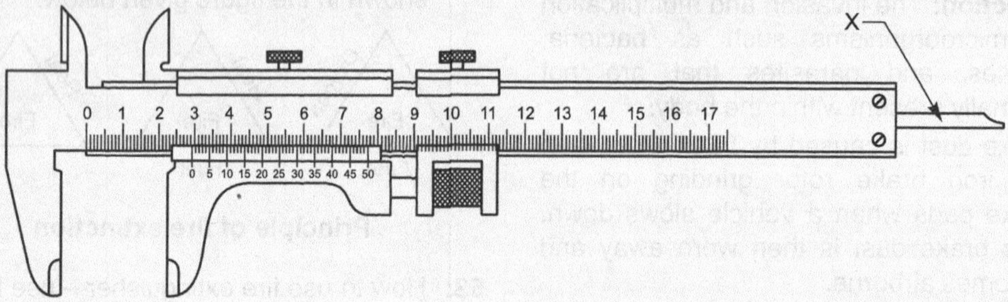

(a) Beam
(b) Fixed jaw
(c) Vernier scale
(d) Depth measuring blade

2. What is the name of the part marked as 'X' in the figure given below?

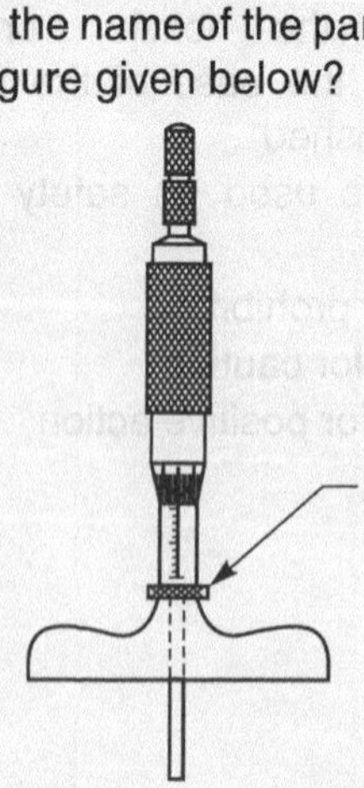

(a) Cap (b) Lock
(c) Stock (d) Graduated sleeve

3. What is the least count of the metric outside micrometer?

(a) 0.01 mm (b) 0.10 mm
(c) 0.0001 mm (d) 0.00001 mm

4. Which part ensures uniform pressure on the measuring faces in the outside micrometer?

(a) Barrel (b) Spindle
(c) Thimble (d) Ratchet stop

5. Which is the fixed measuring face of a micrometer?

(a) Anvil (b) Barrel
(c) Spindle (d) Thimble

6. A tool, which is used for laying out large circles is:

(a) Trammel (b) Divider
(c) Jenny calliper (d) Scriber

7. The point angle of a centre punch is:

(a) 30° (b) 60°
(c) 90° (d) 120°

8. During marking, the reference surface is provided by:
(a) Sketch of the job
(b) Workpiece
(c) Marking off table surface
(d) Surface gauge

9. Identify the tool shown in figure given below.

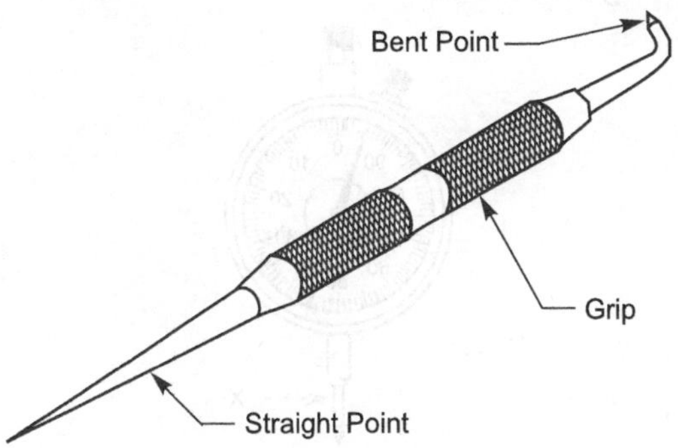

(a) Prick punch
(b) Scriber
(c) Jenny calliper
(d) Drift

10. The jaws in this type of wrench open parallel with the help of a screw thus proving suitable for all sizes of nuts. This is:
(a) Pipe wrench
(b) Monkey wrench
(c) Socket wrench
(d) Allen wrench

11. Which instrument is used to measure the diameter of a camshaft journal?
(a) Depth micrometer
(b) Three-point internal micrometer
(c) Inside micrometer
(d) Outside micrometer

12. Which material used in the outside micrometer anvil?
(a) Brass
(b) Bronze
(c) Carbon
(d) Carbide

13. What is the name of the part marked 'X' in figure given below?

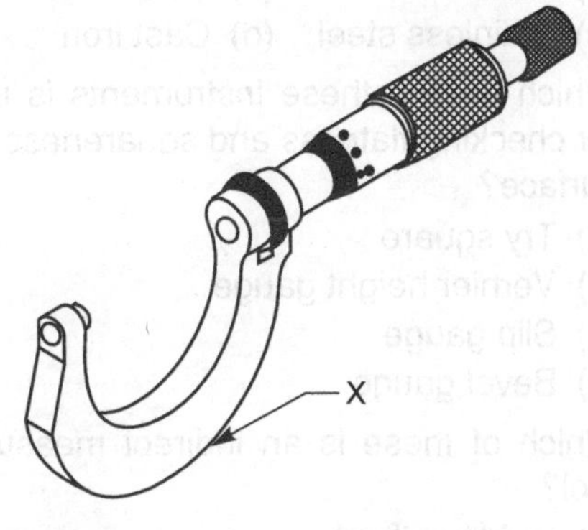

(a) Anvil (b) Frame
(c) Spindle (d) Thimble

14. What is the name of the part marked 'X' in figure given below?

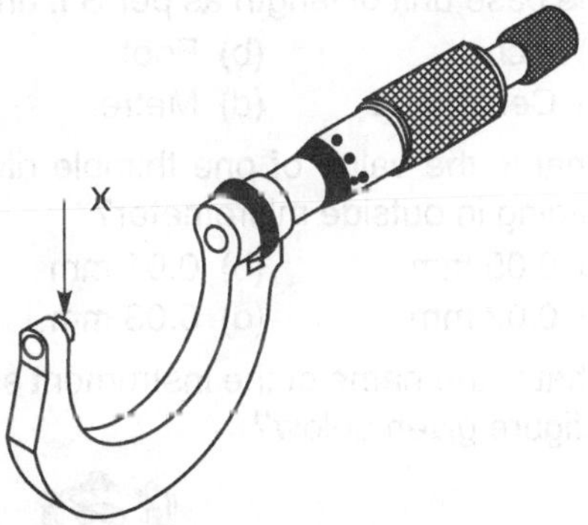

(a) Anvil (b) Frame
(c) Spindle (d) Thimble

15. How many equal thimble divisions are made in outside micrometer?
(a) 30 divisions (b) 40 divisions
(c) 50 divisions (d) 60 divisions

16. Steel rule is a:
(a) Marking instrument
(b) Precision instrument
(c) Checking instrument
(d) Direct reading measuring instrument

17. What is determined while measuring a component?
(a) Nominal size (b) Actual size
(c) Specified size (d) Scale size

18. Steel rule is made of:
(a) Brass (b) Zinc
(c) Stainless steel (d) Cast iron

19. Which among these instruments is used for checking flatness and squareness of a surface?
(a) Try square
(b) Vernier height gauge
(c) Slip gauge
(d) Bevel gauge

20. Which of these is an indirect measuring tool?
(a) Inside calliper
(b) Vernier calliper
(c) Universal bevel protractor
(d) Inside micrometer

21. The base unit of length as per S.I. units is:
(a) Inch (b) Foot
(c) Centimetre (d) Metre

22. What is the value of one thimble division reading in outside micrometer?
(a) 0.05 mm (b) 0.01 mm
(c) 0.02 mm (d) 0.03 mm

23. What is the name of the instrument shown in figure given below?

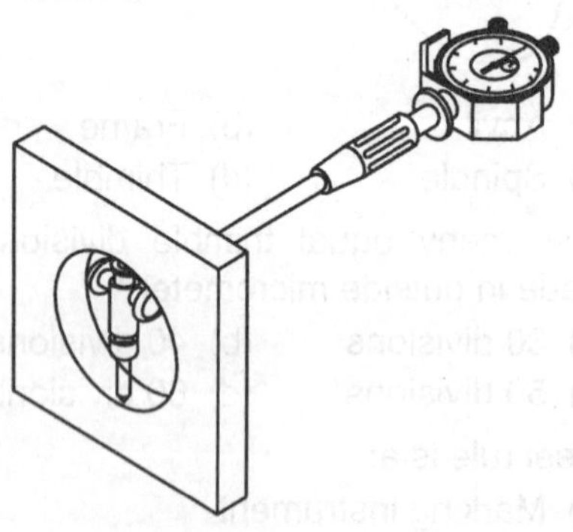

(a) Depth gauge
(b) Bore dial gauge
(c) Telescopic gauge
(d) Vernier gauge

24. Which part actuates the movement of the dial for reading measurement of dial bore gauge?
(a) Stem (b) Plunger
(c) Fixed anvil (d) Centring shoe

25. Which instrument is used to read the telescopic gauge measurement?
(a) Depth micrometer
(b) Inside micrometer
(c) Outside micrometer
(d) Three-point internal micrometer

26. Name the part marked 'X' in the figure given below.

(a) Anvil (b) Stem
(c) Plunger (d) Pointer

27. Name the part marked 'X' in the figure given below.

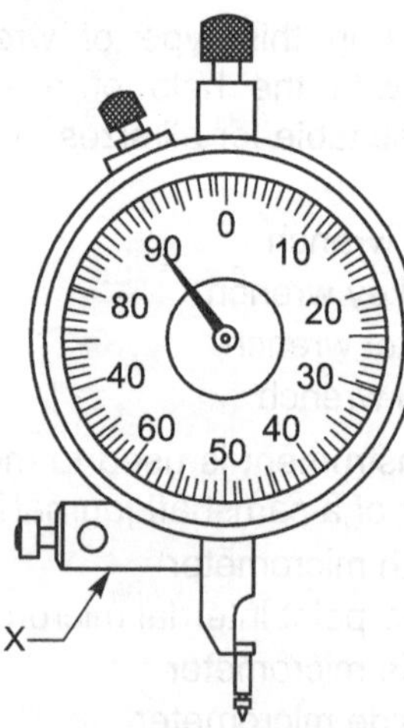

(a) Anvil (b) Clamp
(c) Plunger (d) Stem

28. Which instrument is used to check the end-play of crank shaft?
(a) Dial test indicator
(b) Telescopic gauge
(c) Bore dial gauge
(d) Slip gauge

29. Which one of the following instruments is used to check the concentricity of outside diameter?
(a) Vernier calliper
(b) Outside micrometer
(c) Dial test indicator
(d) Dial calliper

30. Which one of the following instruments is used to measure accurately the angle of a taper?
(a) Bevel protractor
(b) Bevel gauge
(c) Vernier bevel protractor
(d) Taper gauge

31. The least count of vernier calliper is:
(a) 0.10 mm (b) 0.01 mm
(c) 0.05 mm (d) 0.02 mm

32. Which calliper is used to find the centre of a round bar?
(a) Inside calliper (b) Jenny calliper
(c) Vernier calliper (d) Outside calliper

33. Which part of the outside micrometer ensures a uniform pressure between the measuring surfaces?
(a) Anvil (b) Thimble
(c) Spindle lock (d) Ratchet stop

34. Which part of the vernier height gauge is an integral part of the main slide?
(a) Jaw (b) Base
(c) Beam (d) Jaw clamp

35. Which part of the vernier bevel protractor is in contact with the inclined surface while measuring?
(a) Dial (b) Disc
(c) Blade (d) Stock

36. What is the name of the gauge shown in figure given below?

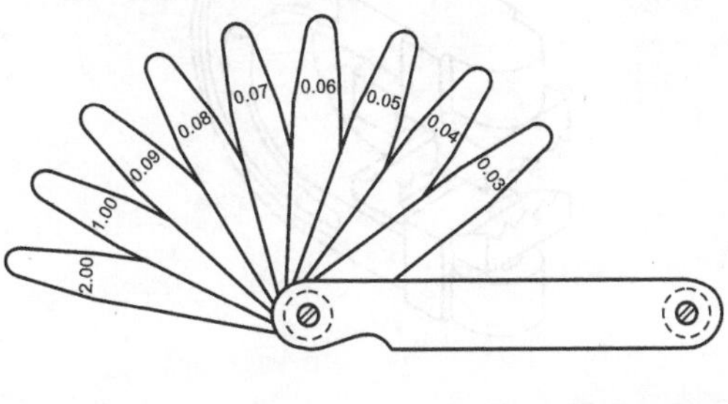

(a) Feeler gauge
(b) Snap gauge
(c) Telescopic gauge
(d) Plug gauge

37. What can be measured by a feeler gauge?
(a) Radius
(b) Screw pitch
(c) Surface roughness
(d) Gap between mating parts

38. Which marking media is used on rough surfaces?
(a) White wash
(b) Prussian blue
(c) Copper sulphate
(d) Cellulose lacquer

39. Which marking media is used to mark on finished surfaces?
(a) White wash
(b) Prussian blue
(c) Copper sulphate
(d) Cellulose lacquer

40. Which marking media is poisonous?
(a) White wash
(b) Prussian blue
(c) Copper sulphate
(d) Cellulose lacquer

41. Which part of combination set is used to mark and check angle of 90° and 45°?
(a) Rule (b) Centre head
(c) Square head (d) Protractor head

42. What is not the use of a try square?
(a) Measuring right angle
(b) Checking squareness
(c) Making straight lines at 90° against an edge
(d) Setting workpiece at 90°

43. The minimum measurement that can be read with the help of a steel rule is:
(a) 0.01 mm (b) 0.02 mm
(c) 0.05 mm (d) 0.50 mm

44. A vernier depth gauge is used for measuring:
(a) External dimensions
(b) Internal dimensions

(c) Step, depth of blind hole
(d) Pitch diameter

45. Which one of the following statements is correct?
(a) Gauges are used to check size
(b) Templates are used to check size
(c) Gauges are used to measure size
(d) Gauges are used to check the shape of components

46. Thread ring gauge is used for checking:
(a) External threads
(b) Internal threads
(c) External diameter of cylindrical jobs
(d) Internal diameter of jobs

47. What is used to protect slip gauges from rust?
(a) Oil (b) Wax
(c) Kerosene (d) Petroleum jelly

48. Which gauge is used to check the accuracy of an internal thread?
(a) Snap gauge
(b) Thread ring gauge
(c) Thread plug gauge
(d) Centre gauge

49. What instrument is shown in figure given below which is used for measuring external diameter?

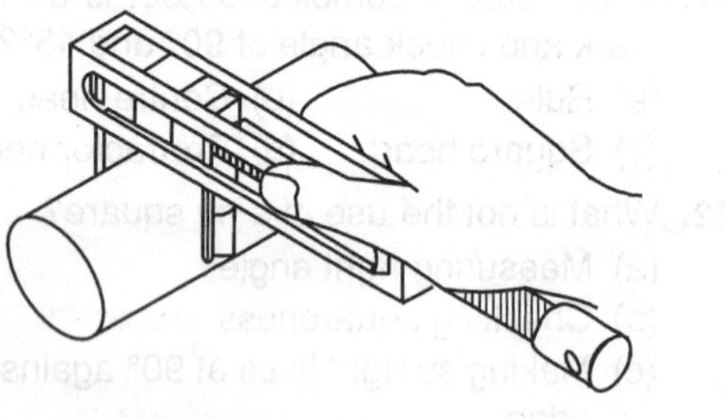

(a) Vernier calliper
(b) Outside calliper
(c) Parallel leg calliper
(d) Pair of special jaws by using slip gauge

50. What device is used to check that a given workpiece conforms to the specified shape?
(a) Profile gauge
(b) Snap gauge
(c) Calliper gauge
(d) Progressive gauge

51. Identify the type of gauge shown in figure given below.

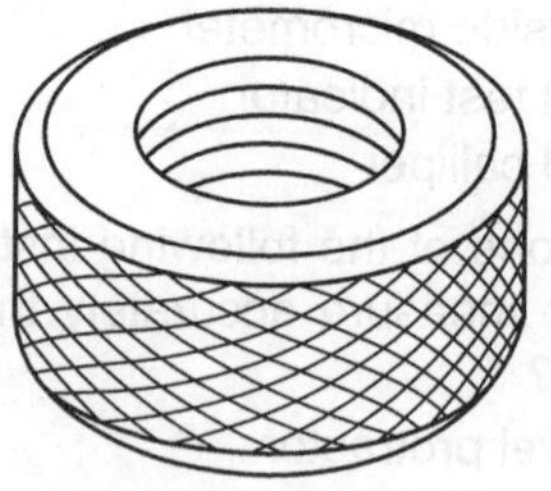

(a) Thread plug gauge
(b) Thread ring gauge
(c) Caliper gauge
(d) Ring gauge

52. Identify the gauge shown in figure given below.

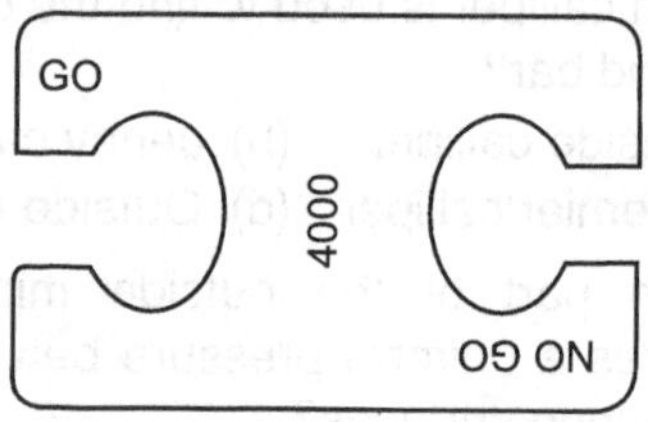

(a) Taper gauge
(b) Snap gauge
(c) Progressive plug gauge
(d) Double-ended plug gauge

53. A drill gauge is used to check:
(a) Diameter of a drill
(b) Length of a drill
(c) Cutting angle of a drill
(d) Lip length of a drill

54. What is the type of gauge shown in figure given below?

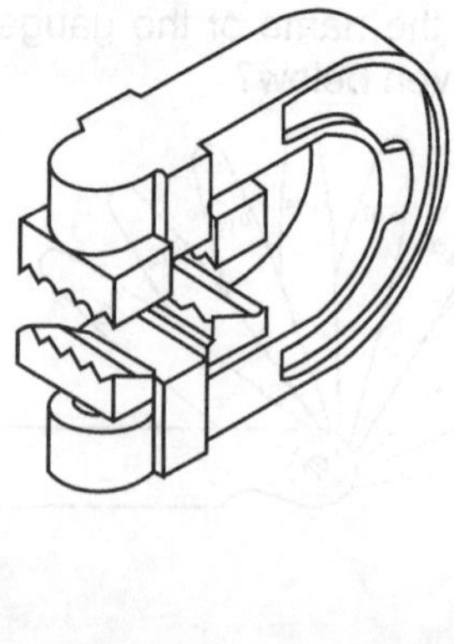

(a) Thread ring gauge
(b) Roller type thread calliper gauge
(c) Thread plug gauge
(d) Anvil type thread snap gauge

55. What is marked as 'X' in the gauge shown in figure given below?

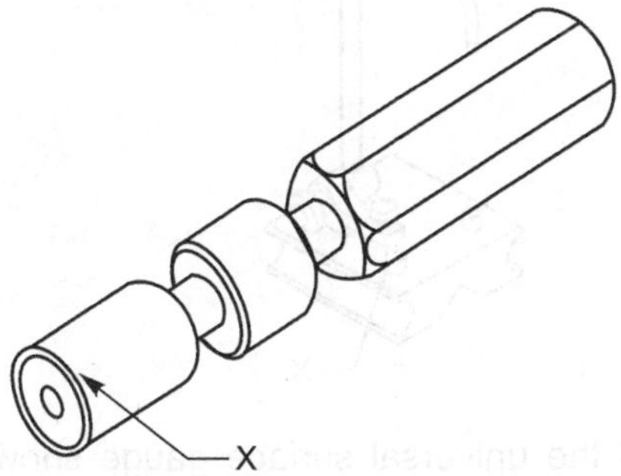

(a) Double-ended plug gauge – NO GO
(b) Progressive plug gauge – GO
(c) Taper plug gauge – GO
(d) Progressive plug gauge – NO GO

56. What is the type of gauge shown in figure given below?

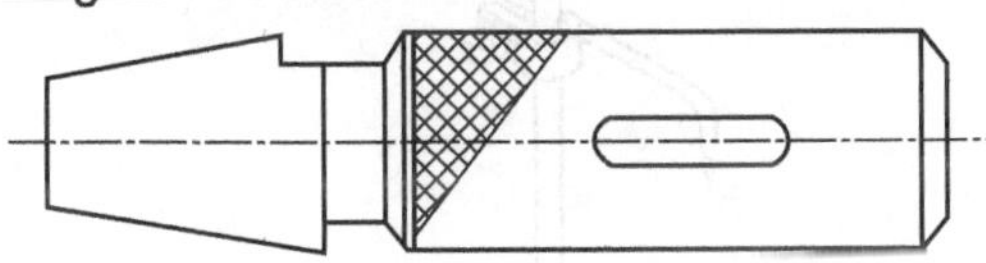

(a) Snap gauge
(b) Progressive plug gauge
(c) Taper ring gauge
(d) Taper plug gauge

57. What is the type of gauge shown in figure given below?

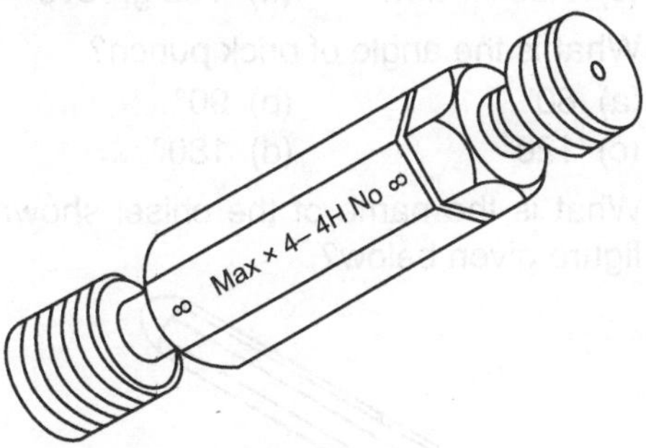

(a) Thread ring gauge
(b) Thread plug gauge
(c) Thread pitch gauge
(d) Taper plug gauge

58. What is the name of the equipment shown in figure given below?

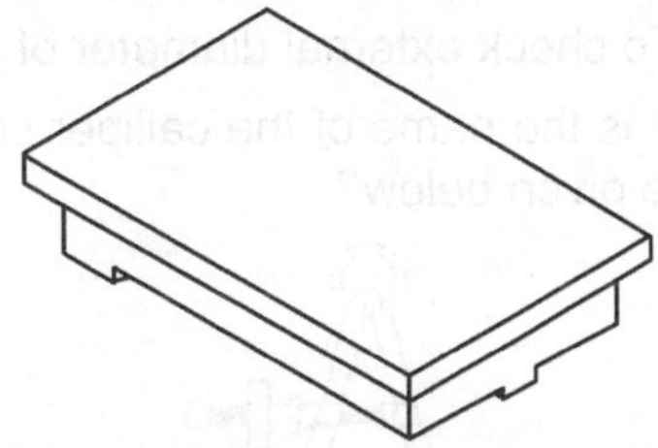

(a) Surface plate (b) Angle plate
(c) 'V' block (d) Parallel block

59. What is the least count of engineer steel rule?

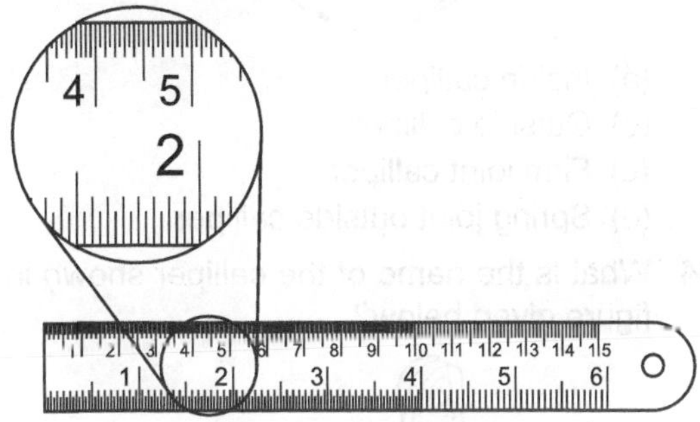

(a) 1 mm (b) 0.5 mm
(c) 2 mm (d) 3 mm

60. Which marking device is used to scribe circles?
(a) Jenny calliper (b) Inside calliper
(c) Spring divider (d) Outside calliper

61. Which instrument is used to check the right angle?
(a) Steel rule
(b) Try square
(c) Firm joint calliper
(d) Spring joint calliper

62. What is the use of outside calliper?

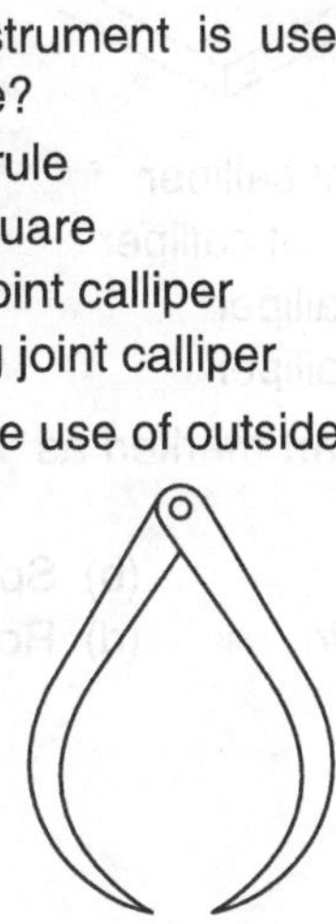

(a) To check flatness

(b) To check an angle
(c) To check internal diameter of a job
(d) To check external diameter of a job

63. What is the name of the calliper shown in figure given below?

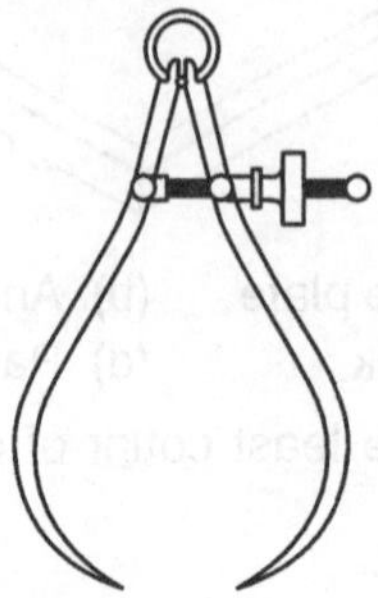

(a) Inside calliper
(b) Outside calliper
(c) Firm joint calliper
(d) Spring joint outside calliper

64. What is the name of the calliper shown in figure given below?

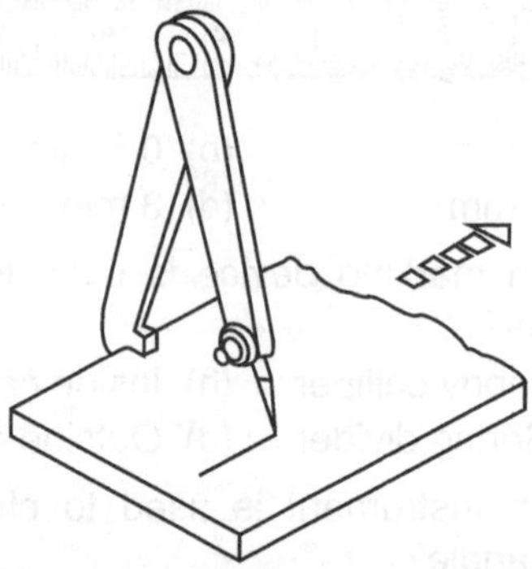

(a) Firm joint calliper
(b) Spring joint calliper
(c) Inside calliper
(d) Jenny calliper

65. Name the part marked as 'X' in the figure given below.

(a) Snug (b) Spindle
(c) Guide pin (d) Rocker arm

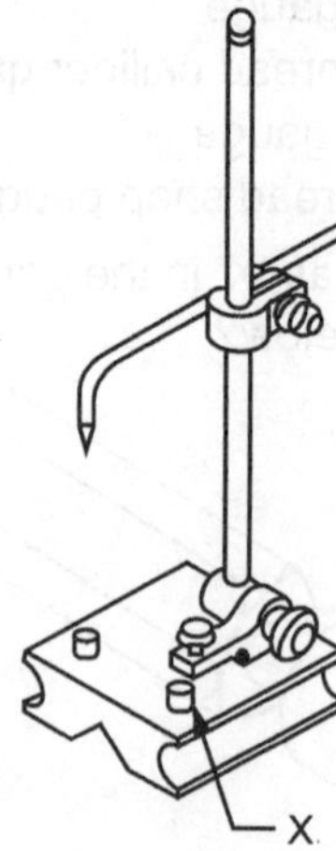

66. In the universal surface gauge shown in figure given below, name the part marked as 'X'.

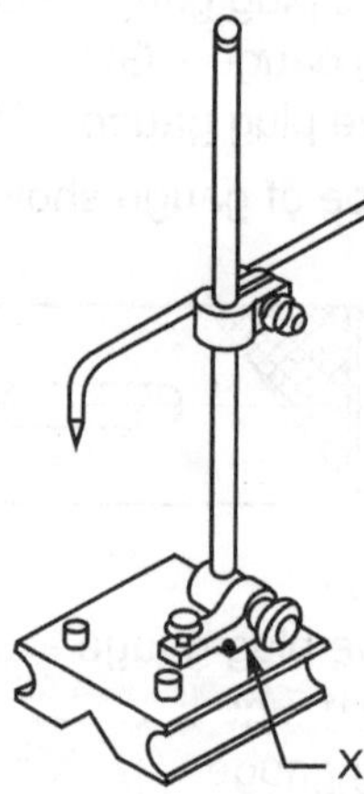

(a) Snug (b) Scriber
(c) Rocker arm (d) Vee groove

67. What is the angle of prick punch?

(a) 60° (b) 90°
(c) 120° (d) 180°

68. What is the name of the chisel shown in figure given below?

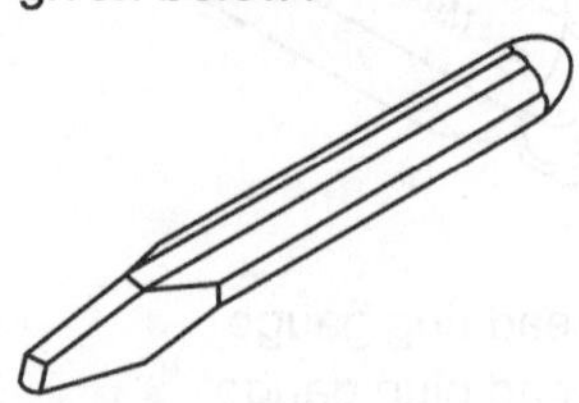

(a) Flat chisel
(b) Cross cut chisel

(c) Half round nose chisel
(d) Diamond point chisel

69. The body of a flat chisel is usually made in the shape of:
(a) Rectangle (b) Square
(c) Octagon (d) Hexagon

70. The chisel used to cut keyways is:
(a) Cape chisel
(b) Flat chisel
(c) Round nose chisel
(d) Diamond point chisel

71. The head of a chisel should not be:
(a) Plain (b) Round
(c) Mushroomed (d) Chamfered

72. A groove is being chipped by a chisel. How will you avoid the metal breaking off as a lump, when reaching the end of the groove?
(a) By chipping with light force when reaching the end
(b) By using solublo oil
(c) By holding the chisel with less inclination
(d) By chipping the end portion from opposite side

73. What is the name of the chisel shown in figure given below?

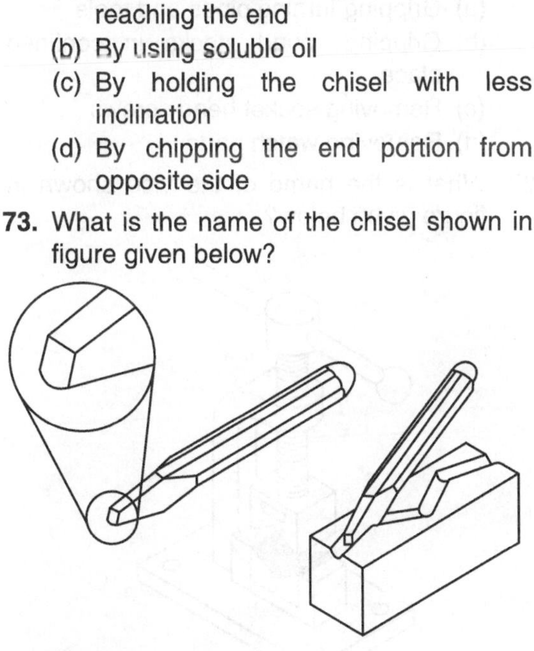

(a) Flat chisel
(b) Cross cut chisel
(c) Half round chisel
(d) Diamond point chisel

74. What is the name of the chisel shown in figure given below?

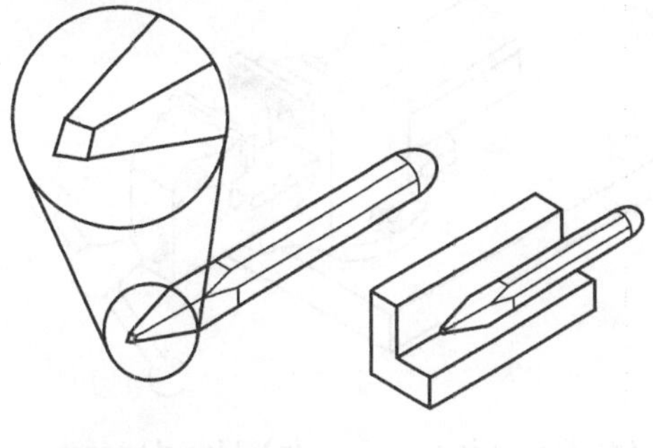

(a) Flat chisel
(b) Cross cut chisel
(c) Half round chisel
(d) Diamond point chisel

75. What is the weight of hammer used for marking purposes?
(a) 200 grams (b) 250 grams
(c) 300 grams (d) 450 grams

76. What is the name of the part of the hammer marked 'X' in figure given below?

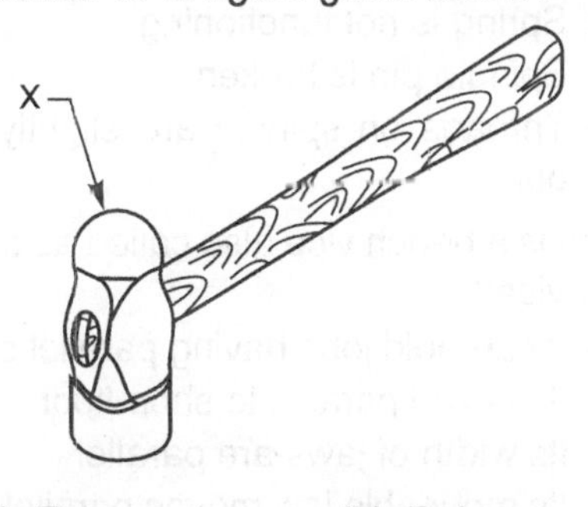

(a) Face (b) Pein
(c) Cheek (d) Eye hole

77. What is the name of the part of the hammer marked 'X' in figure given below?

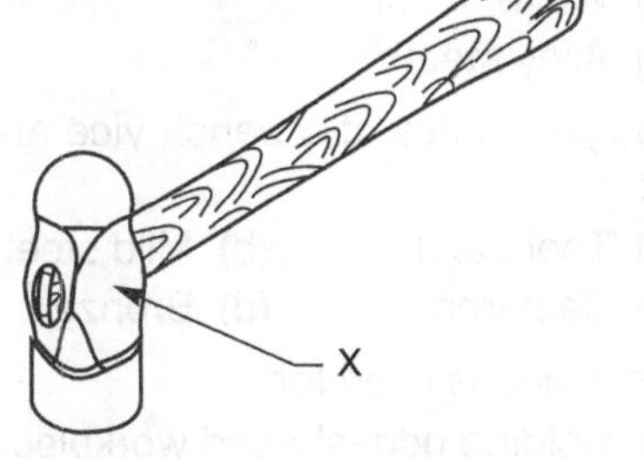

(a) Cheek (b) Wedge
(c) Face (d) Eye hole

78. What is the name of the device shown in figure given below?

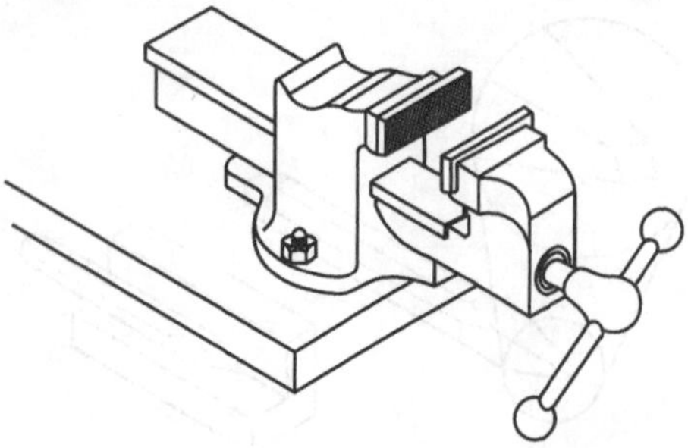

(a) Pipe vice (b) Hand vice
(c) Pin vice (d) Bench vice

79. The best position to hold a job in the vice when filing is:
(a) Eye level (b) Shoulder level
(c) Elbow level (d) Arm level

80. The moveable jaw of a bench vice is not moving even though the spindle is turned. What is the probable reason?
(a) Fixed and movable jaws are over-tight
(b) Spring is not functioning
(c) Spindle pin is broken
(d) Threads on spindle are slightly worn out

81. Why is a bench vice also called as parallel jaw vice?
(a) It can hold jobs having parallel sides
(b) It is fixed parallel to shop floor
(c) Its width of jaws are parallel
(d) Its moveable jaw moves parallel to the fixed jaw

82. The box nut of a bench vice is made of:
(a) Mild steel
(b) Phosphorous bronze
(c) White metal
(d) Alloy steel

83. The jaw plates of a bench vice are made of:
(a) Tool steel (b) Mild steel
(c) Cast iron (d) Bronze

84. A pin vice is used for:
(a) Holding odd–shaped workpieces
(b) Fixing pins
(c) Holding small diameter workpieces
(d) Holding studs

85. A hand vice is used for:
(a) Fixing heavy jobs
(b) Tightening nuts and bolts
(c) Holding round jobs
(d) Doing minute work

86. The spindle of a bench vice is made of:
(a) Mild steel (b) Cast iron
(c) Tool steel (d) Bronze

87. Which type of hammer is used for smoothing and finishing a surface after it has been roughed out to the required shape?
(a) Stretching hammer
(b) Planishing hammer
(c) Cross pein hammer
(d) Engineer's ball pein hammer

88. Allen wrench is used for:
(a) Gripping turning pipes and tools
(b) Gripping round stocks in confined place
(c) Removing socket head screws
(d) Removing watch parts

89. What is the name of the vice shown in figure given below?

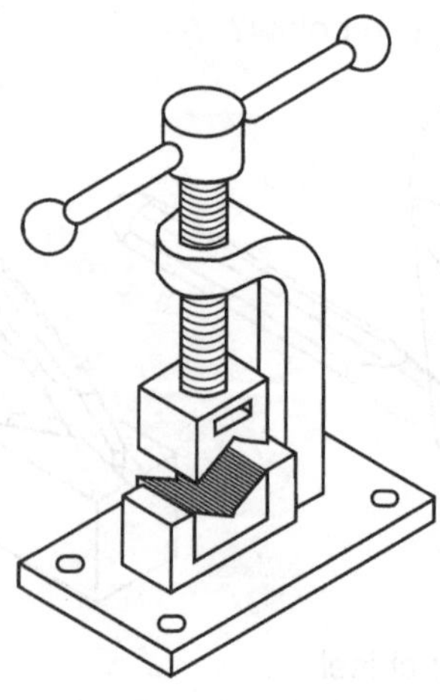

(a) Pipe vice
(b) Hand vice
(c) Tool maker's vice
(d) Bench vice

90. What is the name of the pliers shown in figure given below?

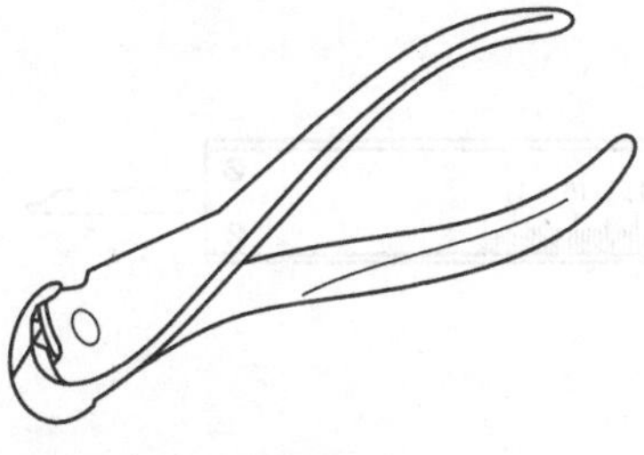

(a) Slip joint pliers
(b) End cutting pliers
(c) Flat nose pliers
(d) Side cutting pliers

91. What is the name of the tool shown in figure given below?

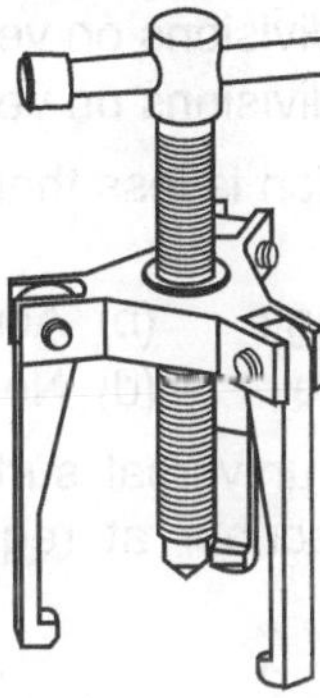

(a) Jack (b) Puller
(c) Wrench (d) Extractor

92. What is the working medium of pneumatic tools?
(a) Oil (b) Air
(c) Water (d) Kerosene

93. What is the name of the screw driver shown in figure given below?

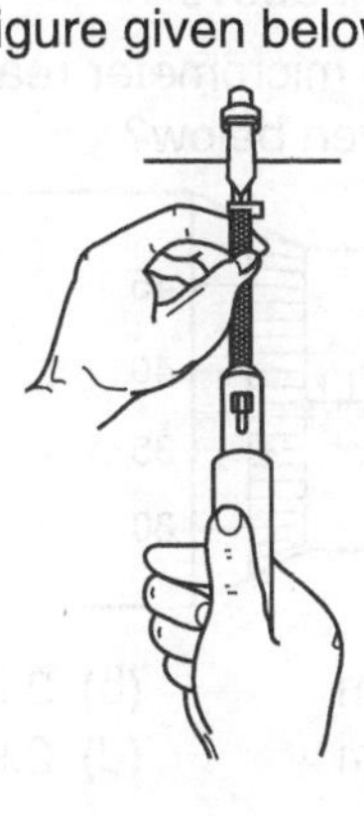

(a) Recess screwdriver
(b) Ratchet screwdriver
(c) Stumpy screwdriver
(d) Offset screwdriver

94. What is the material used to make Allen keys?
(a) Cast iron
(b) Aluminium alloy
(c) High speed steel
(d) Chrome vanadium steel

95. What is the use of a bench vice?
(a) To hold a workpiece
(b) To hold a tool
(c) To hold a work table
(d) To hold a chuck

96. What is the name of the spanner shown in figure given below?

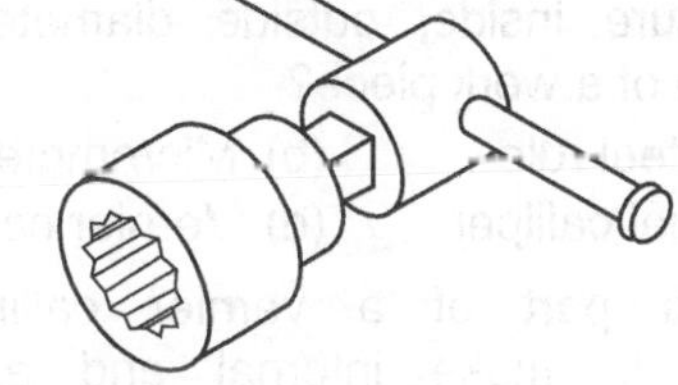

(a) Open spanner
(b) Tubular spanner
(c) Socket spanner
(d) Ring spanner

97. What is the name of the spanner shown in figure given below?

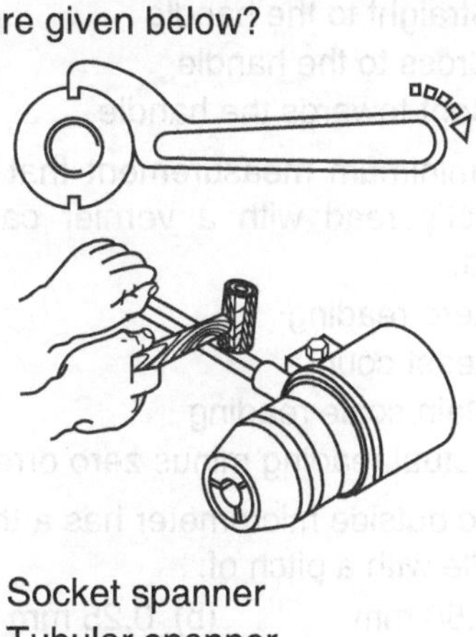

(a) Socket spanner
(b) Tubular spanner
(c) Hook spanner
(d) Adjustable spanner

98. What is the name of the part marked as 'X' in the vernier calliper shown in figure given below?

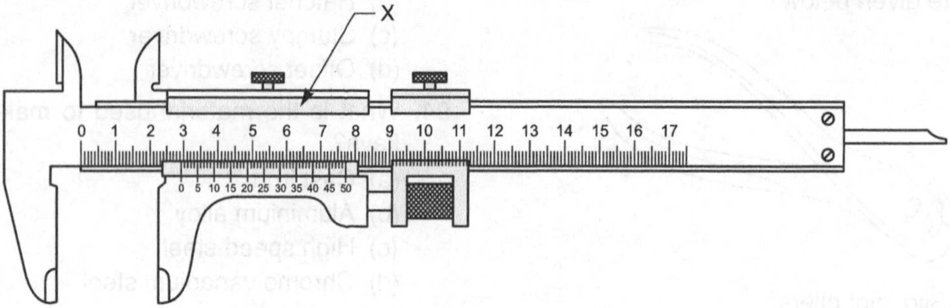

(a) Beam (b) Lock screw
(c) Sliding unit (d) Movable jaw

99. What is the least count of vernier caliper in metric system?
(a) 0.1 mm (b) 0.01 mm
(c) 0.02 mm (d) 0.001 mm

100. Which measuring instrument is used to measure inside, outside diameter and depth of a work piece?
(a) Steel rule (b) Micrometer
(c) Dial calliper (d) Vernier calliper

101. Which part of a vernier calliper is used to make internal and external measurements?
(a) Beam (b) Sliding unit
(c) Fixed jaw (d) Movable jaw

102. The peen of a cross peen hammer is:
(a) Angular to the handle
(b) Straight to the handle
(c) Cross to the handle
(d) Bent towards the handle

103. The minimum measurement that can be correctly read with a vernier calliper is called:
(a) Zero reading
(b) Least count
(c) Main scale reading
(d) Actual reading minus zero error

104. Metric outside micrometer has a threaded spindle with a pitch of:
(a) 0.50 mm (b) 0.25 mm
(c) 1.00 mm (d) 1.50 mm

105. To get the least count of 5′ in a vernier bevel protractor, the 23° main scale are divided into:
(a) 12 equal divisions on vernier scale
(b) 22 equal divisions on vernier scale
(c) 24 equal divisions on vernier scale
(d) 25 equal divisions on vernier scale

106. The angle which is less than 90° is called as:
(a) Obtuse angle (b) Acute angle
(c) Right angle (d) None of these

107. What part of universal surface gauge is used to set scriber at required position exactly?
(a) Snug
(b) Guide pin
(c) Rocker arm
(d) Fine adjustment screw

108. A divider is used for:
(a) Scribing circles
(b) Scribing arcs
(c) Transferring and stepping of distances
(d) All of the above

109. What is the micrometer reading as shown in figure given below?

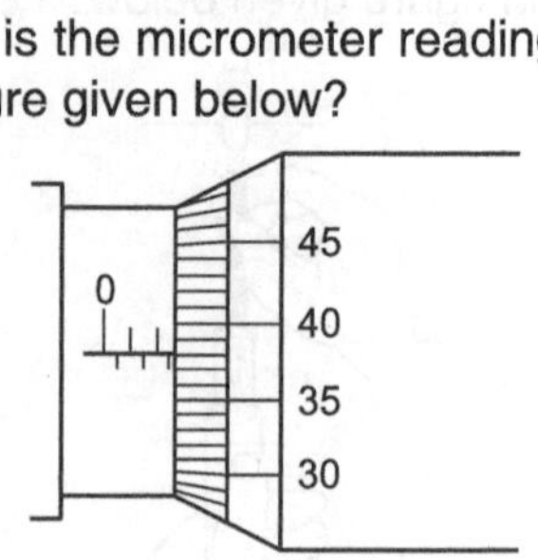

(a) 2.08 mm (b) 2.50 mm
(c) 2.88 mm (d) 2.98 mm

110. Why is a slight convexity made on the cutting edges of flat chisel?
(a) To cut sharp corners
(b) To allow lubricant to reach the cutting edge
(c) To cut curved surfaces
(d) To prevent digging of the ends of the cutting edge

111. Which type of chisel is being used as shown in figure given below?

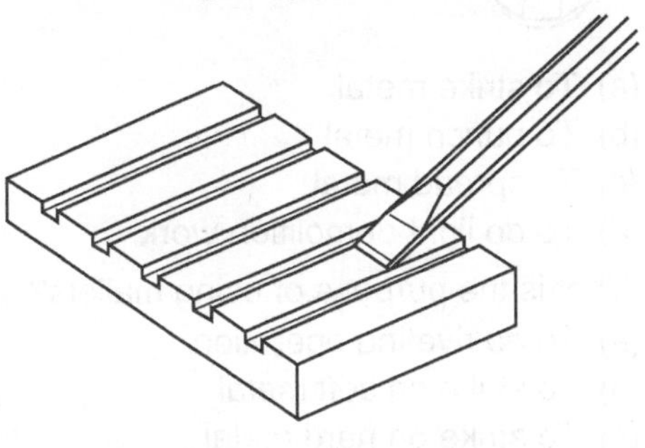

(a) Flat chisel (b) Cross-cut chisel
(c) Half-round chisel (d) Web chisel

112. In the figure given under question 111, if grooves are to be cut, which type of chisel will be used?
(a) Flat chisel
(b) Cross-cut chisel
(c) Half-round chisel
(d) Web chisel

113. Which marking media provides clear lines on machine finished surfaces?
(a) Whitewash
(b) Prussian blue
(c) Copper sulphate
(d) Cellulose lacquer

114. What is the spindle movement of one division of thimble with spindle thread of 0.5 mm pitch?
(a) 0.01 mm (b) 0.10 mm
(c) 1.00 mm (d) 10.0 mm

115. Which calliper provides fine adjustment of dimensions?
(a) Inside calliper
(b) Odd leg calliper
(c) Outside calliper
(d) Spring joint calliper

116. Why are marking tables made of cast iron?
(a) Self-lubricating
(b) Make more rigidity
(c) Maintain the accuracy
(d) Prevent thermal expansion

117. What is the reading of vernier calliper (figure given below) with least count of 0.1 mm?

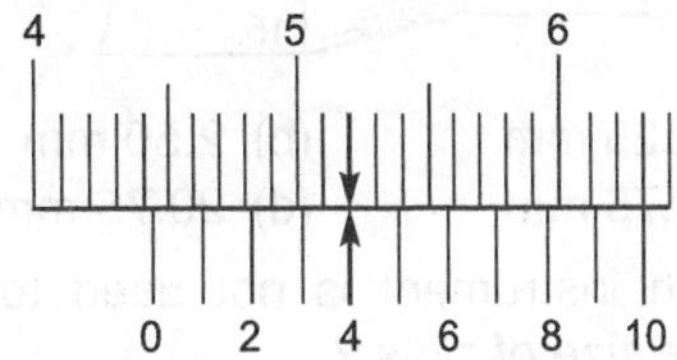

(a) 44.8 mm (b) 44.4 mm
(c) 44.6 mm (d) 45.2 mm

118. What is the reading of vernier calliper (figure given below) with least count of 0.1 mm?

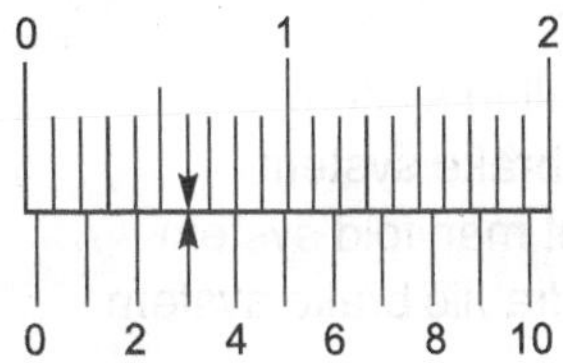

(a) 0.03 mm (b) 0.3 mm
(c) 0.003 mm (d) 1.03 mm

119. What is the smallest possible measurement that can be taken with a depth micrometer?
(a) 0.01 mm (b) 0.02 mm
(c) 0.001 mm (d) 0.002 mm

120. What is the measuring instrument used to measure diameter of a piston pin?
(a) Depth micrometer
(b) Inside micrometer
(c) Outside micrometer
(d) 3-point internal micrometer

121. What is the reading of the outside micrometer (figure below)?

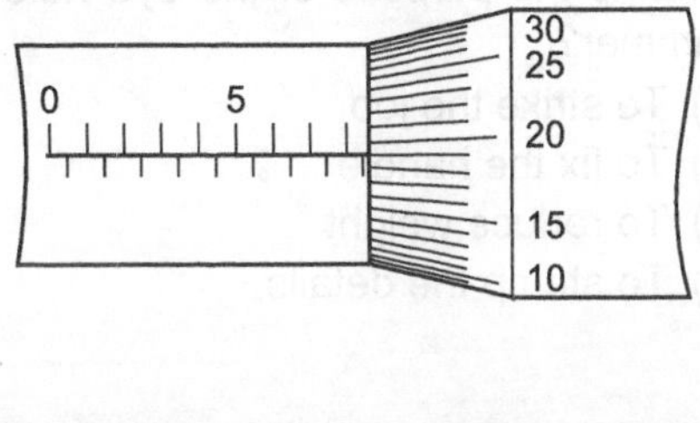

(a) 8.20 mm (b) 8.59 mm
(c) 8.69 mm (d) 8.44 mm

122. What is the reading of the outside micrometer (figure below)?

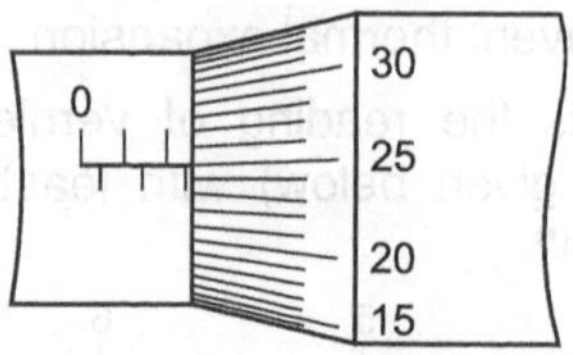

(a) 2.25 mm (b) 2.50 mm
(c) 2.75 mm (d) 20.75 mm

123. Which instrument is not used to check inside size of slots ?
(a) Vernier calliper
(b) Outside micrometer
(c) Inside micrometer
(d) Telescopic gauge

124. A vacuum gauge is used to check leakage in:
(a) Cooling system
(b) Air brake system
(c) Inlet manifold system
(d) Hydraulic brake system

125. What is the purpose of a chisel?
(a) Grinding (b) Machining
(c) Reaming (d) Chipping

126. Which chisel is used to remove metal from large flat surface?
(a) Flat chisel
(b) Cross cut chisel
(c) Half round nose chisel
(d) Diamond point chisel

127. Which type of chisel is used for squaring material at the corners?
(a) Flat chisel
(b) Half round nose chisel
(c) Cross cut chisel
(d) Diamond point chisel

128. What is the purpose of the eye hole in a hammer?
(a) To strike the job
(b) To fix the handle
(c) To reduce weight
(d) To stamp the details

129. What is the use of a lump hammer, shown in figure given below?

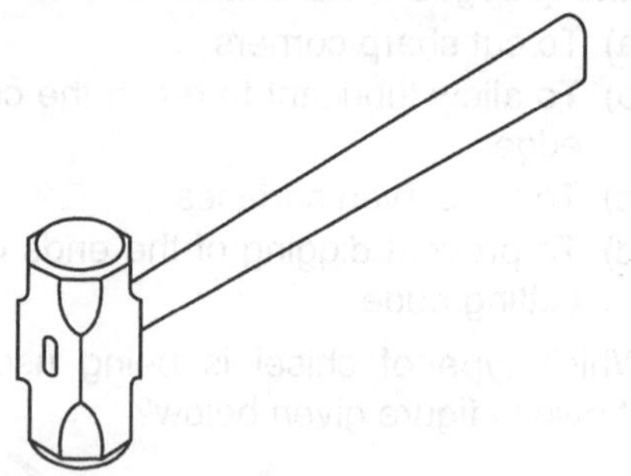

(a) To strike metal
(b) To punch metal
(c) To spread metal
(d) To do light demolition work

130. What is the purpose of using mallets?
(a) To do riveting operation
(b) To strike on soft metal
(c) To strike on hard metal
(d) To spread metal to all sides

131. Which screwdriver is used in limited space?
(a) Cross-recess screwdriver
(b) Ratchet screwdriver
(c) Stumpy screwdriver
(d) Offset screwdriver

132. Which vice is used to hold a very small diameter job?
(a) Pipe vice (b) Hand vice
(c) Pin vice (d) Toolmaker's vice

133. In which condition tyre pressure should not be checked?
(a) When the tyre is cold
(b) When tyre is carrying load
(c) Before start of journey
(d) When tyre is hot after long journey

134. Which one of the following parts of a universal surface gauge helps to draw parallel lines along a datum edge?
(a) Fine adjusting screw
(b) Guide pins
(c) Base
(d) Rocker arm

135. In the figure given below, identify cross peen hammer.

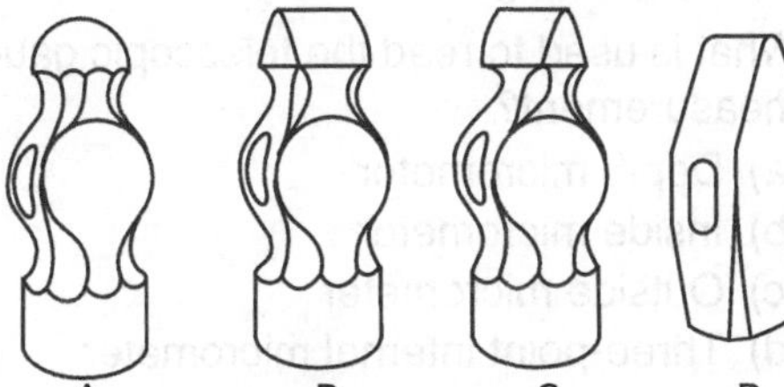

(a) Hammer A (b) Hammer B
(c) Hammer C (d) Hammer D

136. What is the purpose of locknut provided in a micrometer?
(a) To control the movement of the spindle
(b) To lock the reading after setting it over the workpiece
(c) To measure the workpiece accurately
(d) To lock the micrometer when it is not in use

137. What is the principle of micrometer?
(a) Sliding
(b) Screw and nut
(c) Rack and pinion
(d) Worm and worm wheel

138. How can wide range of depth be measured by depth micrometer?
(a) Lengthy sleeve
(b) Lengthy spindle
(c) Adjustable base
(d) Equipped with a set of extension rods

139. Which part indicates the reading of dial test indicator?
(a) Stem (b) Anvil
(c) Pointer (d) Plunger

140. Which chisel is used for separating materials after chain drilling?
(a) Web chisel
(b) Flat chisel
(c) Cross cut chisel
(d) Diamond point chisel

141. Where will be the weight of a hammer stamped?
(a) Face (b) Pein
(c) Cheek (d) Eye hole

142. What is used to finish the re-sharpened divider point?
(a) File (b) Grinder
(c) Scraper (d) Oil stone

143. Which vice is used for holding hollow cylindrical jobs?
(a) Pin vice (b) Pipe vice
(c) Hand vice (d) Bench vice

144. Why is slight convexity given on the face of a hammer?
(a) To make striking effective
(b) To avoid digging of the edge
(c) To allow even loading on striking
(d) To prevent damage to work surface

145. Which scraper is used to scrap the central portion of large flat surface?
(a) Flat scraper
(b) Hook scraper
(c) Triangular scraper
(d) Bull nose scraper

146. Which punch Is used for witness marks?
(a) Dot punch (b) Pin punch
(c) Bell punch (d) Centre punch

147. Which chisel is used for removing materials on the corners of keyways?
(a) Flat chisel
(b) Diamond point chisel
(c) Cross cut chisel
(d) Half round nose chisel

148. Which part is the measuring face fitted to the frame of outside micrometer?
(a) Anvil (b) Barrel
(c) Spindle (d) Ratchet stop

149. Which instrument is used to magnify small variation in sizes by means of pointer on a graduated dial?
(a) Dial test indicator
(b) Inside micrometer
(c) Dial vernier calliper
(d) Vernier micrometer

150. What is termed as the smallest possible measurement in vernier calliper?
(a) Deviation (b) Tolerance
(c) Allowance (d) Least count

151. Which micrometer has provision of interchangeable anvils?
(a) Depth micrometer
(b) Inside micrometer
(c) Outside micrometer
(d) Screw thread micrometer

152. What is used to identify high spots while scraping?
(a) Prussian blue
(b) White chalk
(c) Copper sulphate
(d) Cellulose lacquer

153. How a chipped out scraper can be repaired?
(a) By filing
(b) By lapping
(c) By chiselling
(d) By grinding and honing

154. What is used to read the telescopic gauge measurement?
(a) Depth micrometer
(b) Inside micrometer
(c) Outside micrometer
(d) Three-point internal micrometer

155. If a bench vice is over-tightened, ... is likely to be damaged.
(a) Handle (b) Fixed jaw
(c) Movable jaw (d) Spindle

156. What type of rule is shown in figure given below?
(a) Hinge rule (b) Clip rule
(c) Hook rule (d) Clamp rule

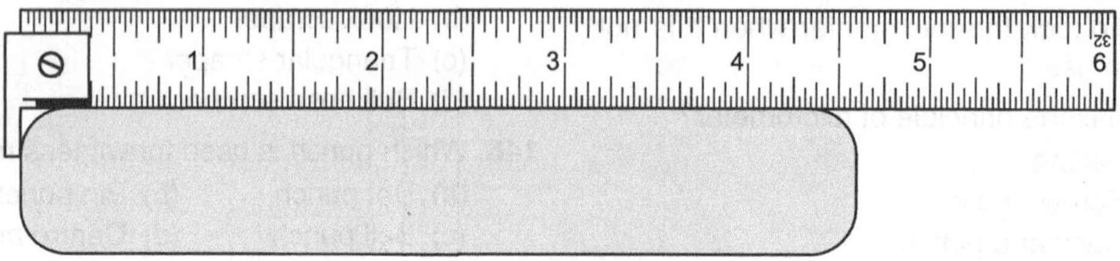

ANSWERS

1. (d)	**2.** (b)	**3.** (a)	**4.** (d)	**5.** (a)	**6.** (a)	**7.** (c)	**8.** (c)	**9.** (b)	**10.** (b)
11. (d)	**12.** (d)	**13.** (b)	**14.** (a)	**15.** (c)	**16.** (d)	**17.** (b)	**18.** (c)	**19.** (a)	**20.** (a)
21. (d)	**22.** (b)	**23.** (b)	**24.** (b)	**25.** (c)	**26.** (c)	**27.** (b)	**28.** (a)	**29.** (c)	**30.** (c)
31. (d)	**32.** (b)	**33.** (d)	**34.** (a)	**35.** (c)	**36.** (a)	**37.** (d)	**38.** (a)	**39.** (b)	**40.** (c)
41. (c)	**42.** (a)	**43.** (d)	**44.** (c)	**45.** (a)	**46.** (a)	**47.** (d)	**48.** (c)	**49.** (d)	**50.** (a)
51. (b)	**52.** (b)	**53.** (a)	**54.** (d)	**55.** (b)	**56.** (b)	**57.** (b)	**58.** (a)	**59.** (b)	**60.** (c)
61. (b)	**62.** (d)	**63.** (d)	**64.** (d)	**65.** (c)	**66.** (c)	**67.** (a)	**68.** (b)	**69.** (c)	**70.** (a)
71. (c)	**72.** (d)	**73.** (c)	**74.** (d)	**75.** (b)	**76.** (b)	**77.** (a)	**78.** (d)	**79.** (c)	**80.** (c)
81. (d)	**82.** (b)	**83.** (a)	**84.** (c)	**85.** (d)	**86.** (a)	**87.** (b)	**88.** (c)	**89.** (a)	**90.** (b)
91. (b)	**92.** (b)	**93.** (b)	**94.** (d)	**95.** (a)	**96.** (c)	**97.** (c)	**98.** (c)	**99.** (c)	**100.** (d)
101. (d)	**102.** (c)	**103.** (b)	**104.** (a)	**105.** (a)	**106.** (b)	**107.** (d)	**108.** (d)	**109.** (c)	**110.** (d)
111. (a)	**112.** (b)	**113.** (b)	**114.** (a)	**115.** (d)	**116.** (a)	**117.** (b)	**118.** (b)	**119.** (a)	**120.** (c)
121. (c)	**122.** (c)	**123.** (b)	**124.** (c)	**125.** (d)	**126.** (a)	**127.** (d)	**128.** (b)	**129.** (d)	**130.** (b)
131. (c)	**132.** (c)	**133.** (d)	**134.** (b)	**135.** (c)	**136.** (b)	**137.** (b)	**138.** (d)	**139.** (c)	**140.** (a)
141. (c)	**142.** (d)	**143.** (b)	**144.** (c)	**145.** (a)	**146.** (a)	**147.** (b)	**148.** (a)	**149.** (a)	**150.** (d)
151. (d)	**152.** (a)	**153.** (d)	**154.** (c)	**155.** (d)	**156.** (c)				

EXPLANATORY NOTES

3: The smallest value that can be measured by the measuring instrument is called its least count.

25: The figure given below shows how an outside micrometer is used to read measurement by a telescopic gauge.

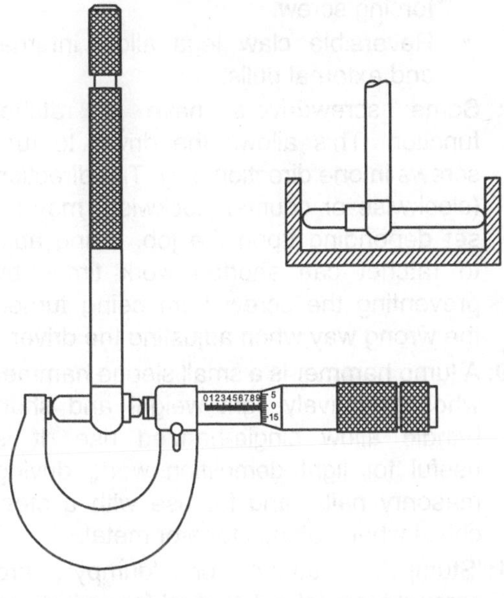

Use of outside micrometer to read telescopic gauge

26: Dial indicators are measuring devices designed expressly to measure relative position. The primary parts of a dial indicator are the face or dial, the case, and the plunger. The plunger is a spring loaded part that can be depressed into the case causing the dial to move clockwise.

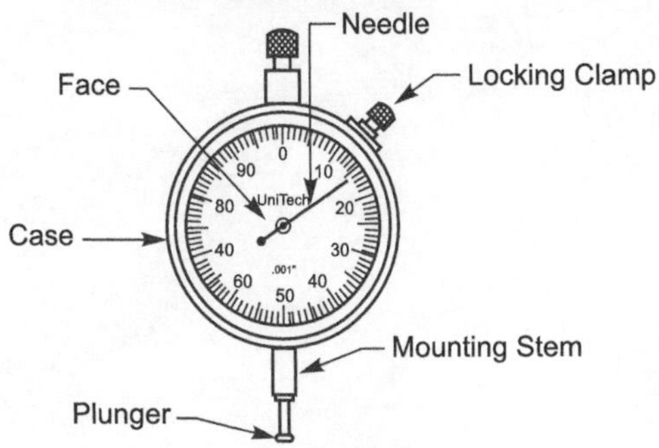

Dial indicator

58: When accurate dimensional features are to be marked, it is essential to have a datum plane with a perfectly flat surface. Marking using datum surfaces which are not perfectly flat will result in dimensional inaccuracies. The most widely used datum surface is the surface plate. More details of surface plate are shown in figure given below.

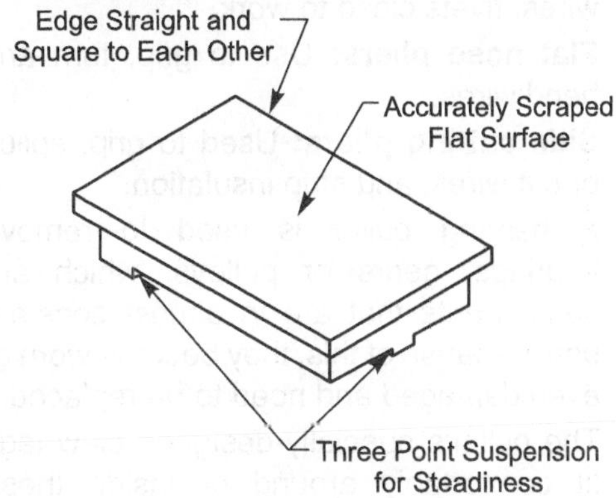

Surface plate

65: Different parts of a surface gauge are shown in figure given below.

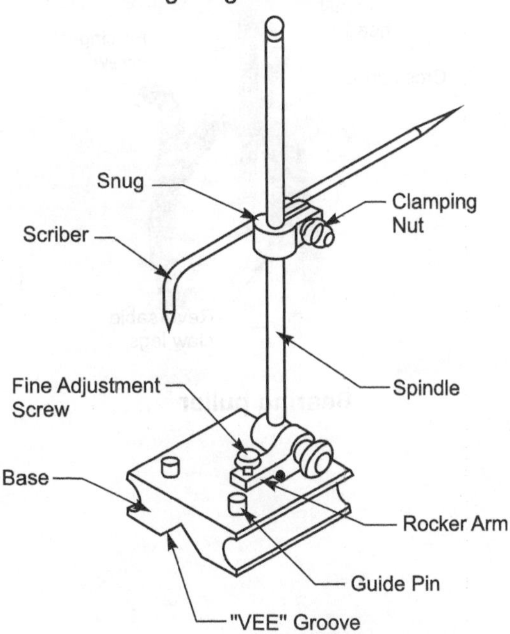

Parts of surface gauge

84: A pin vice is shown in figure given below.

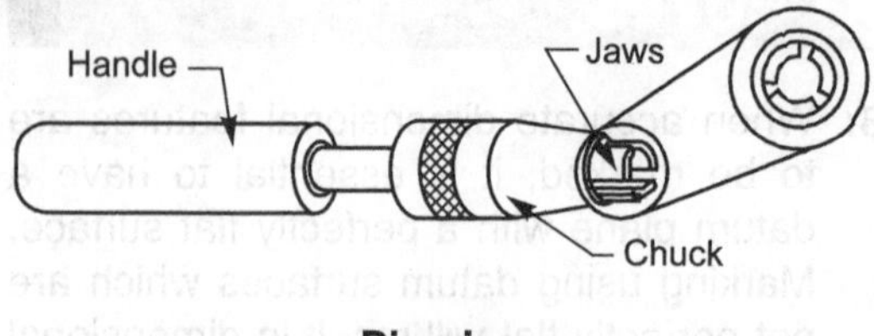

Pin vice

90: **Slip-joint pliers:** Used to adjust nuts and bolts.

End cutting pliers: Used to cut nails, wires, rivets close to work.

Flat nose pliers: Use to grip, turn and bend wires.

Side cutting pliers: Used to grip, splice or cut wires, and strip insulation.

91: A bearing puller is used to remove bearings, gears or pulleys, which are components that are in almost constant use. Because of this, they become worn or even damaged and need to be replaced.

The puller's specially-designed claw legs fit comfortably around or inside these components and a forcing screw applies force to make removal simple, with minimal effort for the user.

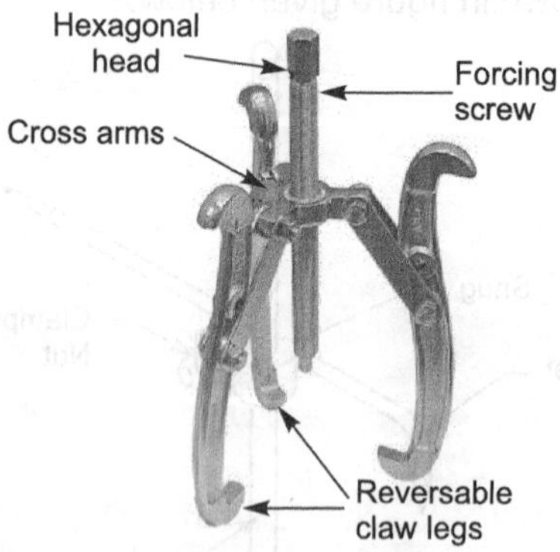

Bearing puller

- Hexagonal head is attached to the forcing screw. It is easily rotated with a socket wrench or spanner.
- Forcing screw is a long, fine threaded screw that is rotated against the centre of the shaft. When the forcing screw is rotated, it applies a force to the component you are removing.
- Cross arm attaches the legs to the forcing screw.
- Reversible claw legs allow internal and external pulls.

93: Some screwdrivers have a ratchet function. This allows the driver to turn screws in one direction only. The direction (clockwise or counter-clockwise) may be set depending upon the job. Being able to ratchet can shorten work times by preventing the screw from being turned the wrong way when adjusting the driver.

129: A lump hammer is a small sledge hammer whose relatively light weight and short handle allow single-handed use. It is useful for light demolition work, driving masonry nails, and for use with a steel chisel when cutting stone or metal.

131: 'Stumpy', 'stubby' or 'dumpy' are screwdrivers (about 40 mm) for use where there is limited space.

Module 3

Fastening and Fitting

MULTIPLE CHOICE QUESTIONS

1. What is the name of the screw shown in figure given below?

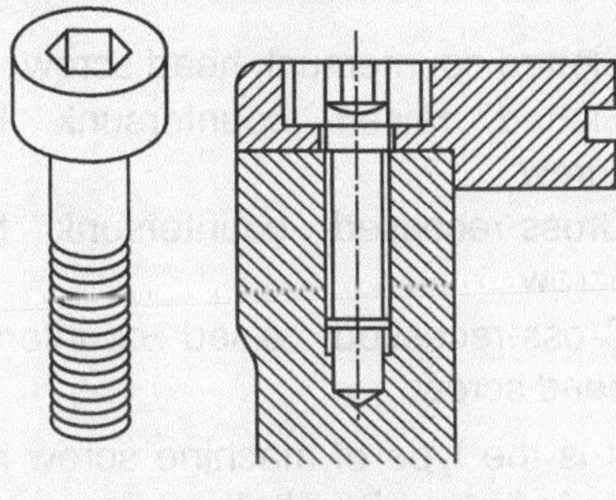

(a) Hoxagon head screw
(b) Hexagon socket head screw
(c) Square head countersunk head screw
(d) Thumb screw

2. What is the name of the washer shown in figure given below?

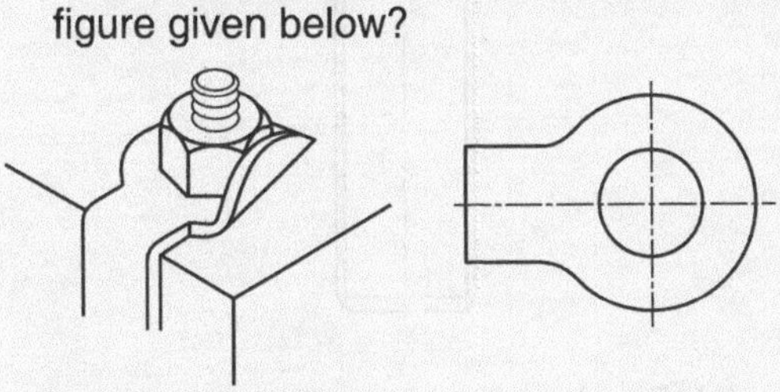

(a) Plain washer
(b) Tab washer
(c) Spring washer
(d) Toothed lock washer

3. What is the name of the screw shown in figure given below?

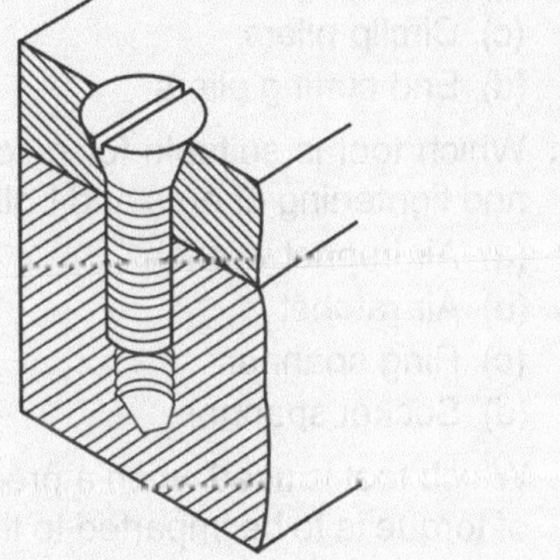

(a) Grub screw
(b) Thumb screw
(c) Hexagonal screw
(d) Countersunk head screw

4. What is the name of the screw shown in figure given below?

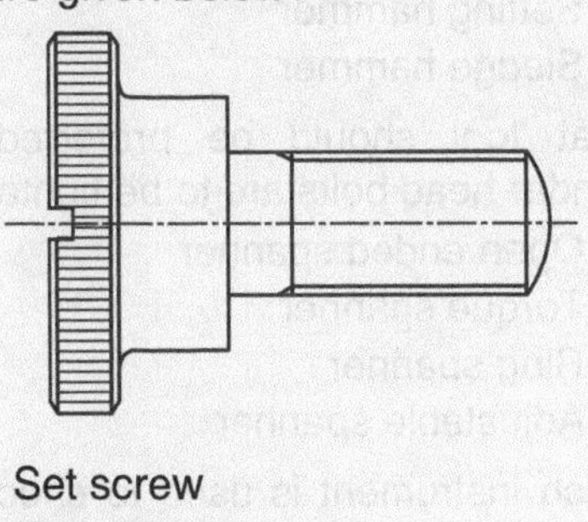

(a) Set screw
(b) Thumb screw
(c) Hexagonal screw
(d) Grub screw

5. What is the name of the pliers shown in figure given below?

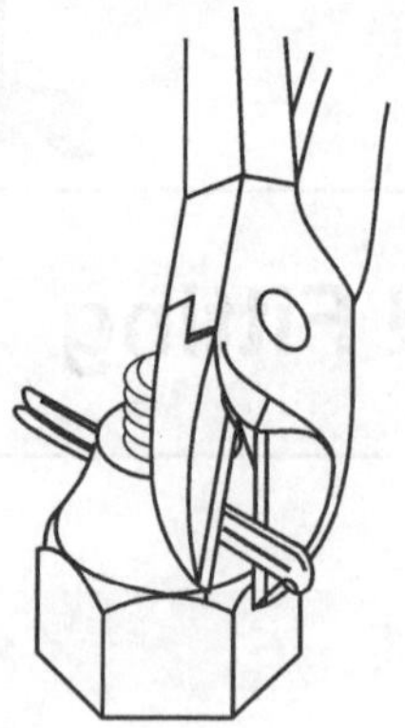

(a) Side cutting pliers
(b) Nose pliers
(c) Circlip pliers
(d) End cutting pliers

6. Which tool is suitable for quick loosening and tightening of nuts and bolts?
(a) Air impact wrench
(b) Air ratchet
(c) Ring spanner
(d) Socket spanner

7. Which tool is used when a precise amount of torque is to be imparted to the fastener?
(a) Air ratchet (b) Ring spanner
(c) Torque wrench (d) Monkey wrench

8. The tool used to flatten the metal around a punched hole is ....
(a) Ball peen hammer
(b) Riveting hammer
(c) Setting hammer
(d) Sledge hammer

9. What tool should be preferred when cylinder head bolts are to be tightened?
(a) Open ended spanner
(b) Torque spanner
(c) Ring spanner
(d) Adjustable spanner

10. Which instrument is used to check taper and ovality of crankshaft journal?
(a) Inside calliper (b) Micrometer
(c) Bore gauge (d) Feeler gauge

11. What is the name of the countersunk screw shown in figure given below?

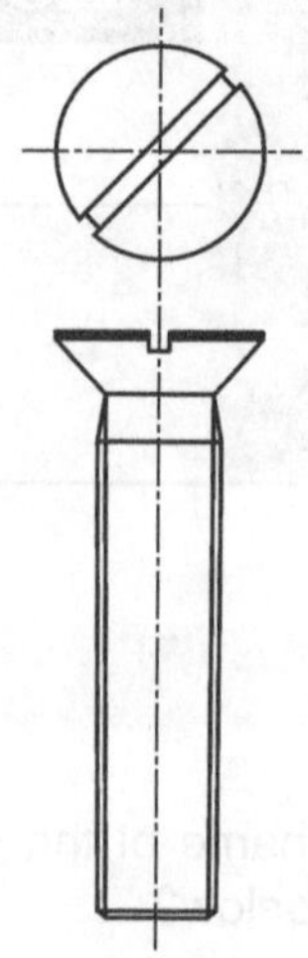

(a) Slotted countersunk head screw
(b) Slotted raised countersunk head screw
(c) Cross-recessed countersunk head screw
(d) Cross-recessed raised countersunk head screw

12. What is the type of machine screw head shown in figure given below?

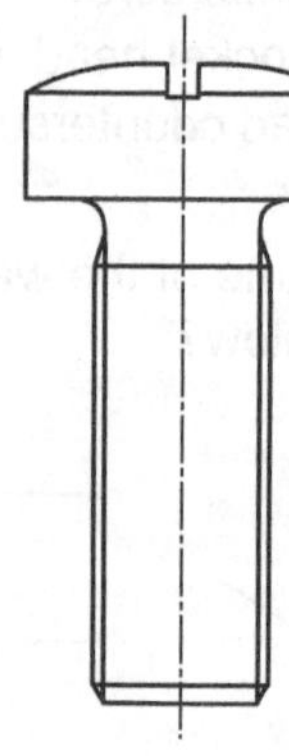

(a) Pan head
(b) Round head
(c) Cheese head
(d) Raised cheese head

13. What is the name of the nut shown in figure given below?
(a) Square nut (b) Lock nut
(c) Castle nut (d) Self locking nut

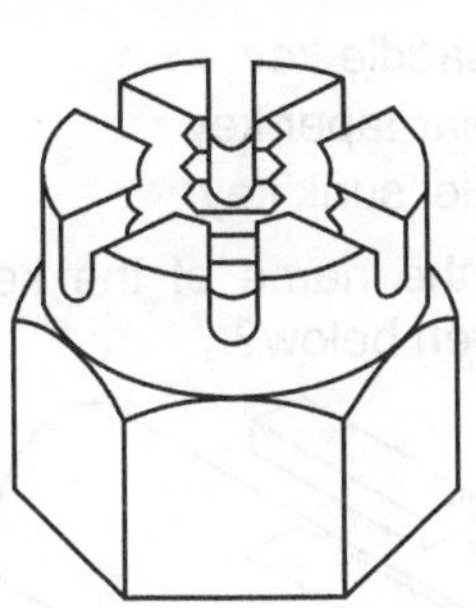

14. What is the name of the nut shown in figure given below?

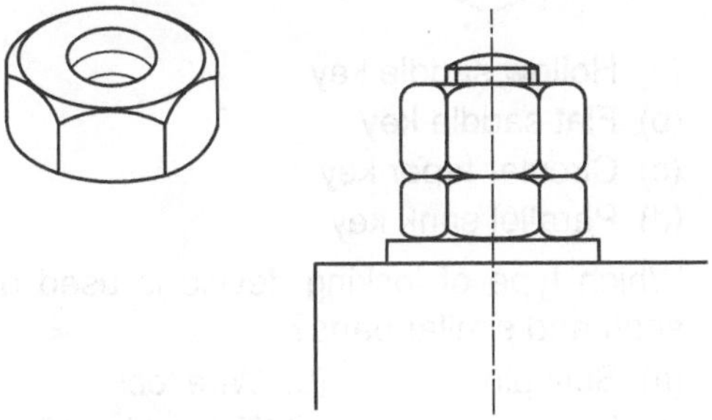

(a) Lock nut
(b) Castle nut
(c) Square nut
(d) Slotted round nut

15. Which type of nut is used for structural and machine tool construction?

(a) Square nut (b) Hexagonal nut
(c) Wing nut (d) Castle nut

16. What is the name of the tool shown in figure given below?

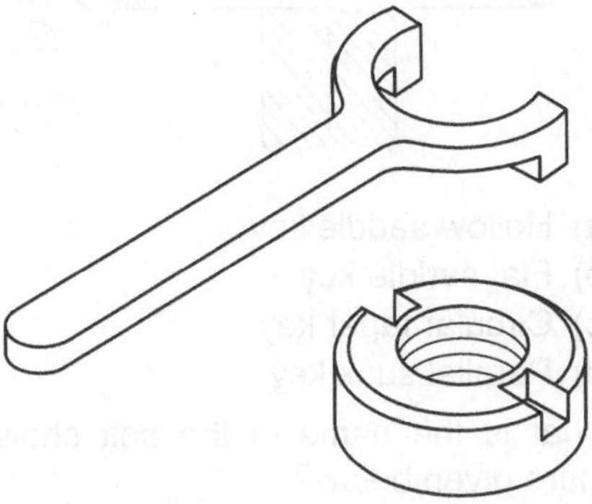

(a) Face wrench (b) Hook wrench
(c) Pin wrench (d) Face pin wrench

17. What is the name of the tool shown in figure given below?

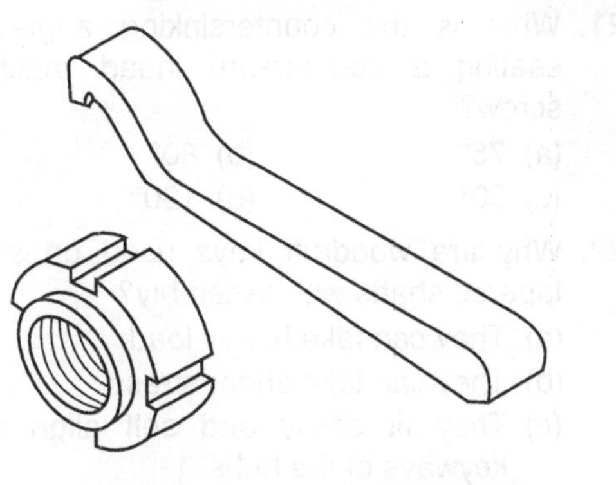

(a) Face wrench (b) Hook wrench
(c) Pin wrench (d) Face pin wrench

18. What is the name of the nut shown in figure given below?

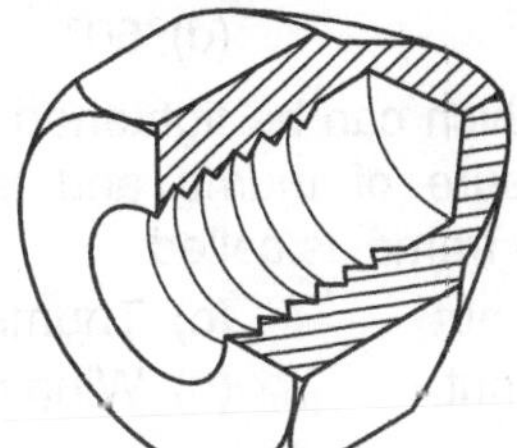

(a) Lock nut
(b) Thumb nut
(c) Domed cap nut
(d) Hexagonal nut with collar

19. What is the name of the nut shown in figure given below?

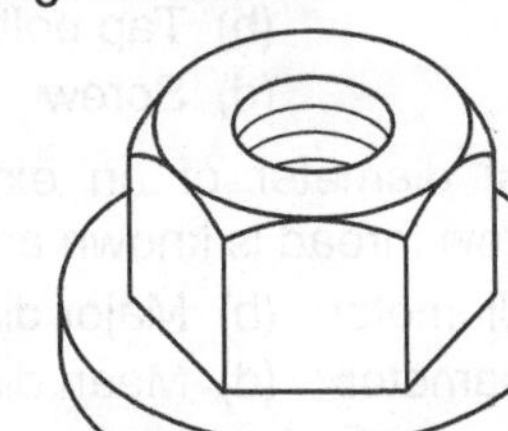

(a) Thumb nut
(b) Domed cap nut
(c) Hexagonal nut with collar
(d) Chuck nut

20. Which type of nut is required to align the axle shaft hole and slot of the nut for locking?

(a) Castle nut (b) Chuck nut
(c) Thumb nut (d) Self lock nut

21. What is the countersinking angle for seating a countersunk head machine screw?
(a) 75° (b) 80°
(c) 90° (d) 120°

22. Why are woodruff keys used on small tapered shafts with assembly?
(a) They can take heavy load
(b) They can take shear stress
(c) They fit easily and self align with keyways of the hubs
(d) They are less expensive

23. For chipping mild steel, the cutting angle of the chisel is:
(a) 45° (b) 50°
(c) 55° (d) 60°

24. A nut, which can be tightened / loosened by pressure of thumb and is used on hacksaw frame, is called:
(a) Ring nut (b) Thumb nut
(c) Cap nut (d) Wing nut

25. The eyehole of a hammer is made oval and tapering towards the centre so as to:
(a) Reduce its weight
(b) Make it easy to manufacture
(c) Reduce effort while using it
(d) Wedge the handle

26. Which of these has threads on both sides?
(a) Bolt (b) Tap bolt
(c) Stud (d) Screw

27. The largest diameter of an external or internal screw thread is known as:
(a) Minor diameter (b) Major diameter
(c) Pitch diameter (d) Mean diameter

28. What is the name of the key shown in figure given below?

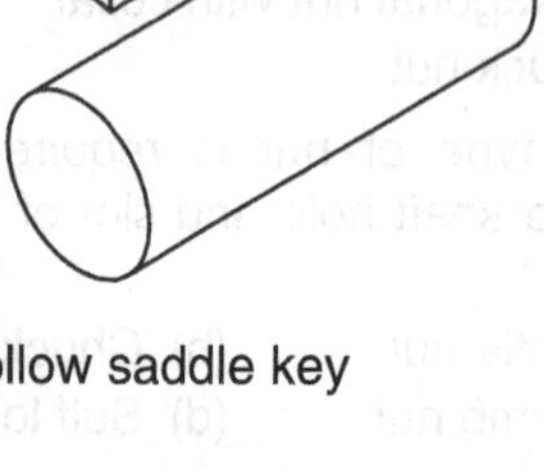

(a) Hollow saddle key
(b) Flat saddle key
(c) Circular taper key
(d) Parallel sunk key

29. What is the name of the key shown in figure given below?

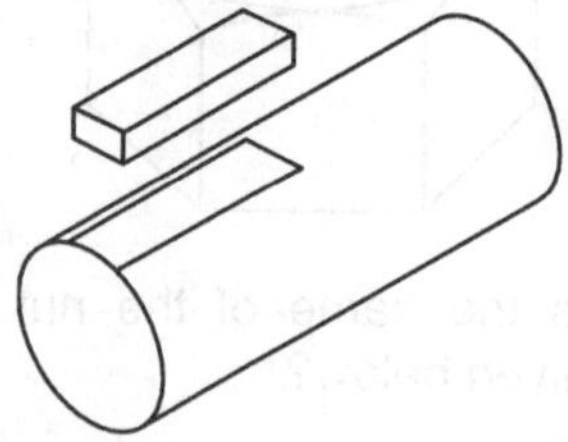

(a) Hollow saddle key
(b) Flat saddle key
(c) Circular taper key
(d) Parallel sunk key

30. Which type of locking device is used on shaft and similar parts?
(a) Split pin (b) Wire lock
(c) Internal circlip (d) External circlip

31. What is the type of key used on the shaft and hub, as shown in the figure given below?

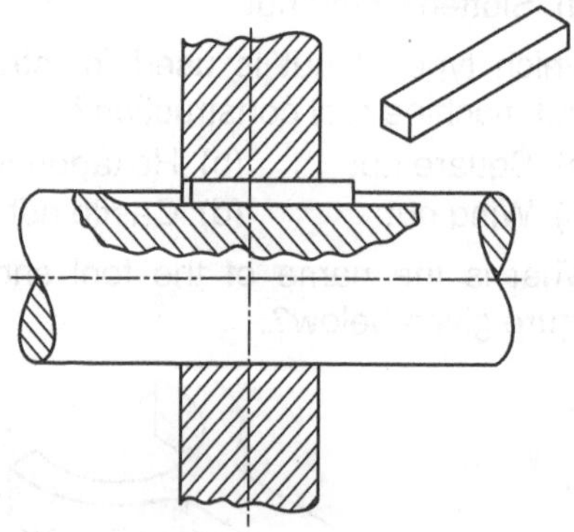

(a) Hollow saddle key
(b) Flat saddle key
(c) Circular taper key
(d) Parallel sunk key

32. What is the name of the bolt shown in figure given below?
(a) Anti-fatigue bolt
(b) Bolt with clearance hole
(c) 'T' bolt
(d) Body fit bolt

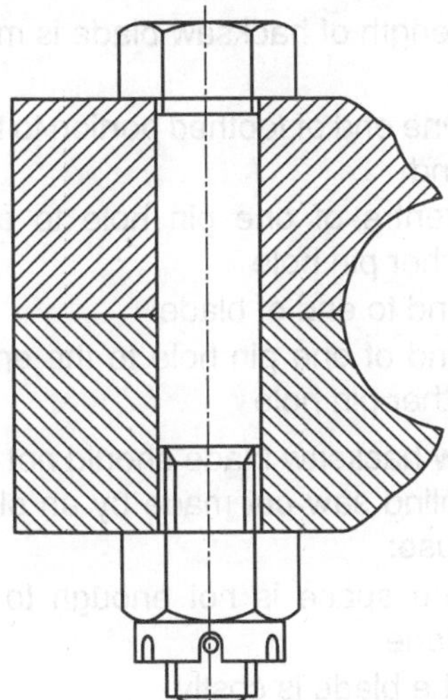

33. What is the name of the key shown in figure given below?

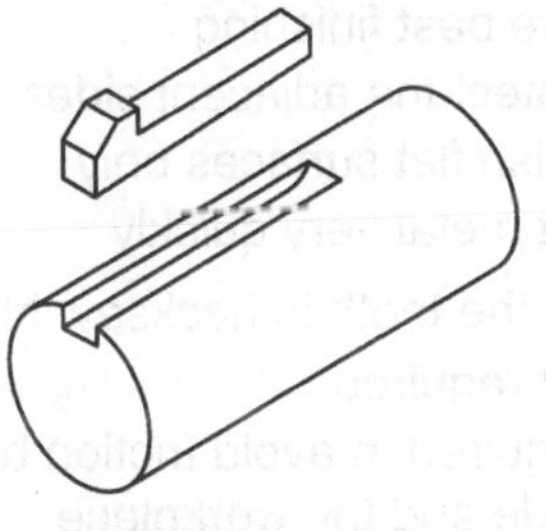

(a) Circular taper key
(b) Parallel sunk key
(c) Gib head key
(d) Woodruff key

34. The distance a drill moves into the work for each revolution of the spindle is called:
(a) Feed of drill
(b) Speed of drill
(c) Depth of drill
(d) Cutting speed of drill

35. A file card is used for:
(a) Cleaning the workpiece
(b) Cleaning teeth of a file
(c) Renewing the teeth of a file
(d) Cleaning the chips

36. A cutting tool used to finish and enlarge a drilled hole is known as:
(a) Tap (b) Die
(c) Reamer (d) Borer

37. The grooves provided on the entire length of the body of a twist drill are called:
(a) Lips (b) Flutes
(c) Margins (d) Webs

38. The point angle of a twist drill used depends upon:
(a) Cutting speed
(b) Type of drilling machine
(c) Size of the drill
(d) Material to be drilled

39. What drilling machine can be used if source of electric power is not available?
(a) Bench drilling machine
(b) Pillar drilling machine
(c) Radial drilling machine
(d) Ratchet drilling machine

40. The device which can be used to hold and locate a round piece for drilling is:
(a) Adjustable locator
(b) 'V' block
(c) Pins locator
(d) Wedge type locator

41. What method of metal cutting operation is shown in figure given below?

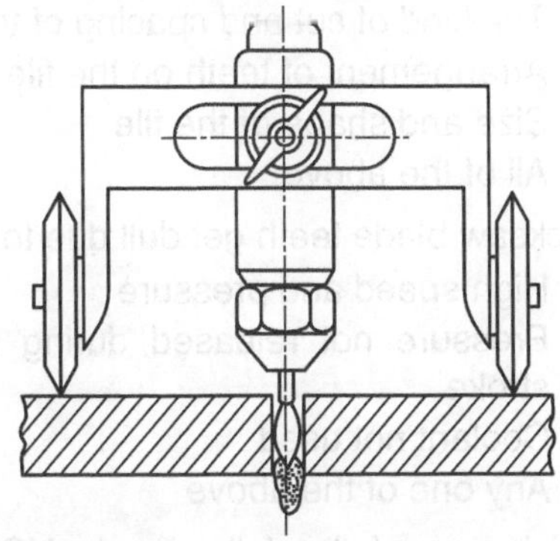

(a) Abrasion (b) Fusion
(c) Incision (d) Rough grinding

42. What is the name of the file shown in figure given below?

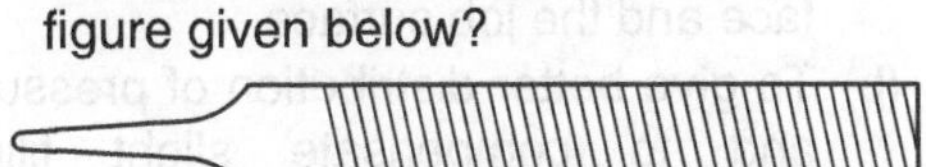

(a) Single cut (b) Double cut
(c) Rasp cut (d) Curved cut

43. What is the name of the file shown in figure given below?

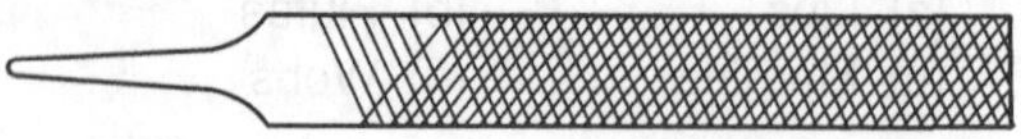

(a) Single cut (b) Double cut
(c) Rasp cut (d) Curved cut

44. What is the name of the file shown in figure given below?

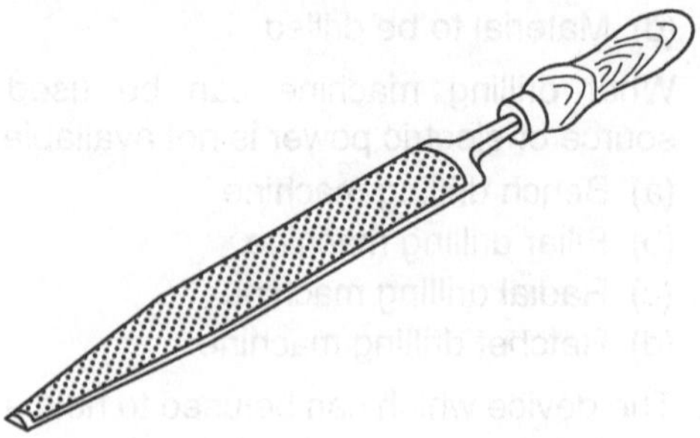

(a) Single cut (b) Double cut
(c) Rasp cut (d) Curved cut

45. What is called metal worker's pencil?
(a) Scriber (b) Chisel
(c) Punch (d) Divider

46. The cutting action of a file depends upon:
(a) The kind of cut and spacing of teeth
(b) Arrangement of teeth on the file
(c) Size and shape of the file
(d) All of the above

47. Hacksaw blade teeth get dull due to:
(a) High speed and pressure
(b) Pressure not released during return stroke
(c) Coolant not used
(d) Any one of the above

48. Which one of the following is NOT the reason for making file faces slightly convex in form?
(a) To increase friction between the file face and the job surface
(b) To give better distribution of pressure and to compensate slight filing movement
(c) To reduce friction between the file face and the job surface
(d) To reduce effort in driving a file

49. The length of hacksaw blade is measured from:
(a) One end of toothed portion to the other end
(b) Centre of one pin hole to centre of other pin hole
(c) End to end of blade
(d) End of one pin hole to the end of the other pin hole

50. A new hacksaw blade should not be used in a blind saw cut made by an old blade, because:
(a) The space is not enough to run the blade
(b) The blade is costly
(c) The new blade has sharp teeth
(d) None of these

51. The safe edge of a file is used to:
(a) Give best finishing
(b) Protect the adjacent sides
(c) Finish flat surfaces only
(d) Cut metal very quickly

52. Setting the teeth in hacksaw blade is:
(a) Not required
(b) Required to avoid friction between the blade and the workpiece
(c) Required for increasing the strength of the blade
(d) Required for providing teeth on the blade

53. The most suitable pitch of the hacksaw blade for cutting thin section tubes is:
(a) 0.8 mm (b) 1.0 mm
(c) 1.4 mm (d) 1.8 mm

54. The most suitable pitch of the hacksaw for cutting solid brass is:
(a) 0.8 mm (b) 1.0 mm
(c) 1.4 mm (d) 1.8 mm

55. The commonly used standard length of hacksaw blade is:
(a) 100 mm (b) 150 mm
(c) 200 mm (d) 250 mm

56. A warding file is similar to:
(a) Flat file but uniform in width
(b) Hand file but narrow in width

(c) Flat file but thinner in thickness
(d) Flat file but thinner and uniform in thickness

57. In case of a single cut file, the teeth are cut on the file face at an angle of:
(a) 45° to the centre line of the file
(b) 60° to the centre line of the file
(c) 75° to the centre line of the file
(d) 80° to the centre line of the file

58. What is the use of fine finish hacksaw blades?
(a) To cut soft metal
(b) To cut high speed steel
(c) To cut high carbon steel
(d) To cut conduit and other thin tubes

59. What type of hacksaw frame is shown in figure given below?

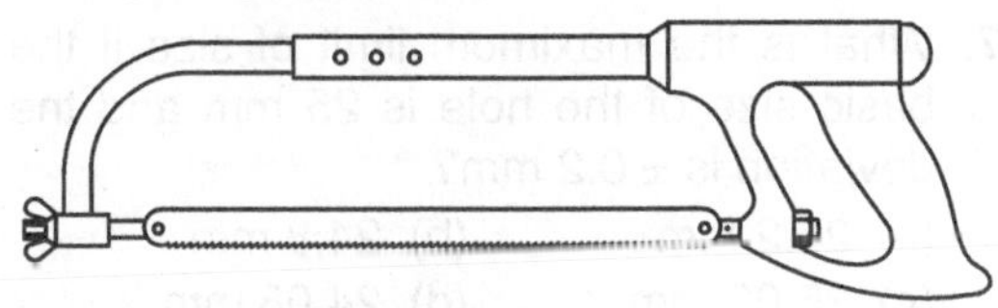

(a) Deep cutting frame
(b) Adjustable flat type frame
(c) Adjustable tubular frame
(d) Solid frame

60. Name the method of removing broken stud as shown in figure given below?

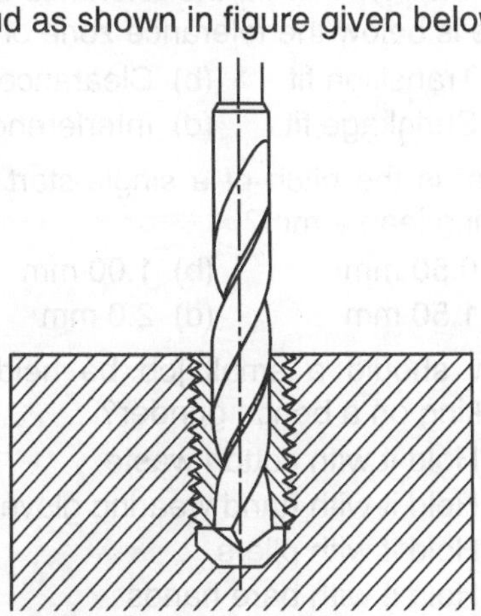

(a) Prick punch method
(b) Filling square form method
(c) Making drill hole method
(d) Ezy-out method

61. Name the method of removing broken stud as shown in figure given below?

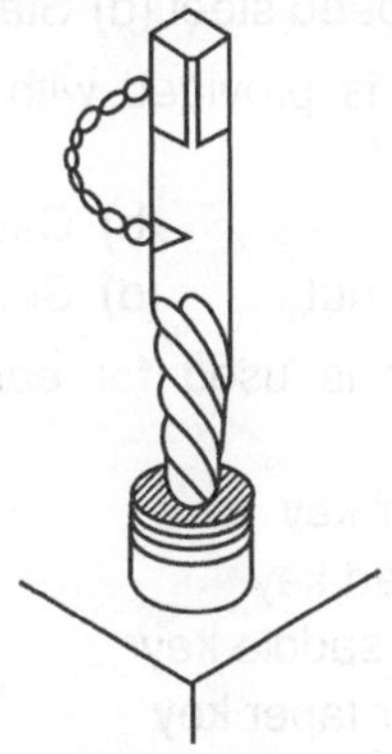

(a) Making drill hole method
(b) Ezy-out method
(c) Filling square form method
(d) Prick punch method

62. What is the name of the drilling machine shown in figure given below?

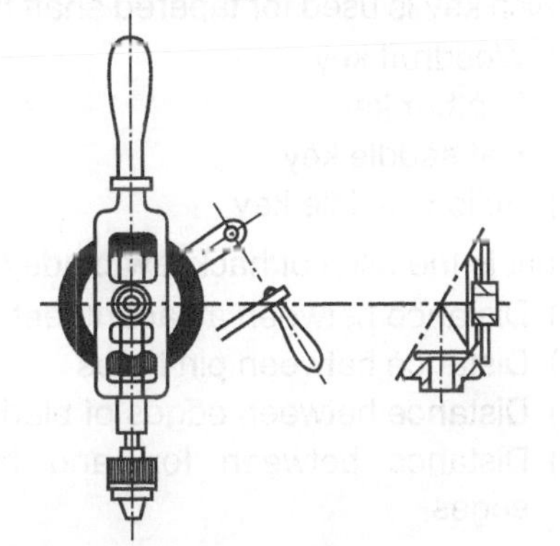

(a) Ratchet drilling machine
(b) Bevel gear drilling machine
(c) Pneumatic drilling machine
(d) Breast drilling machine

63. Which type of portable drilling machine has variable speed?
(a) Breast hand drill
(b) Heavy duty electric hand drill
(c) Ratchet hand drill
(d) Bevel gear hand drill

64. Which part of drill bit fits into the slot of drilling machine spindle?
(a) Body (b) Tang
(c) Flute (d) Web

65. Hand taps are made of:
(a) Cast iron (b) Low carbon steel
(c) High speed steel (d) Stainless steel

66. Which nut is provided with nylon ring in the groove?
(a) T – nut (b) Castle nut
(c) Slotted nut (d) Self-locking nut

67. Which key is used for easy fixing and removal?
(a) Feather key
(b) Gib head key
(c) Hollow saddle key
(d) Circular taper key

68. Which key is used while hub slides axially on the shaft?
(a) Woodruff key
(b) Gib – head key
(c) Feather key
(d) Hollow saddle key

69. Which key is used for tapered shaft fitting?
(a) Woodruff key
(b) Feather key
(c) Flat saddle key
(d) Hollow saddle key

70. What is the pitch of hacksaw blade?
(a) Distance between adjacent teeth
(b) Distance between pin holes
(c) Distance between edges of blade
(d) Distance between top and bottom edges

71. Which type of file is used for filing wood and leather?
(a) Rasp cut file (b) Bastard file
(c) Second cut file (d) Smooth file

72. What is meant by 'tolerance' in the 'limits and fits' system?
(a) Difference between actual size and basic size
(b) Difference between standard and minimum size
(c) Difference between standard and maximum size
(d) Difference between maximum and minimum size

73. Which type of fit will have maximum size hole and minimum size shaft?
(a) Interference fit (b) Clearance fit
(c) Transition fit (d) Shrinkage fit

74. Which type of fit represents the tolerance zone of the hole and shaft overlapping each other?
(a) Transition fit (b) Clearance fit
(c) Shrinkage fit (d) Interference fit

75. What is the term used for the relationship that exists between two mating parts?
(a) Fit (b) Limit
(c) Tolerance (d) Allowance

76. What is the lower limit of size if dimension is stated as $ø\ 25^{+0.021}_{-0.000}$ mm?
(a) 24.85 mm (b) 24.75 mm
(c) 25.00 mm (d) 25.021 mm

77. What is the maximum limit of size if the basic size of the hole is 25 mm and the deviation is ± 0.2 mm?
(a) 25.2 mm (b) 24.8 mm
(c) 25.02 mm (d) 24.08 mm

78. What is the maximum limit of size if size on drawing is $20^{-0.12}_{-0.21}$ mm?
(a) 19.988 mm (b) 19.979 mm
(c) 20.012 mm (d) 20.021 mm

79. In which type of fit, the tolerance zone of hole is below the tolerance zone of shaft?
(a) Transition fit (b) Clearance fit
(c) Shrinkage fit (d) Interference fit

80. What is the pitch of a single-start thread having lead 2 mm?
(a) 0.50 mm (b) 1.00 mm
(c) 1.50 mm (d) 2.0 mm

81. How should a small job be held while working on a bench grinder?
(a) Hold it with cotton waste
(b) Hold it with hand wearing gloves
(c) Hold it with pliers
(d) Hold it with bare hands

82. What operation is being done on the grinding wheel as shown in the figure given below?

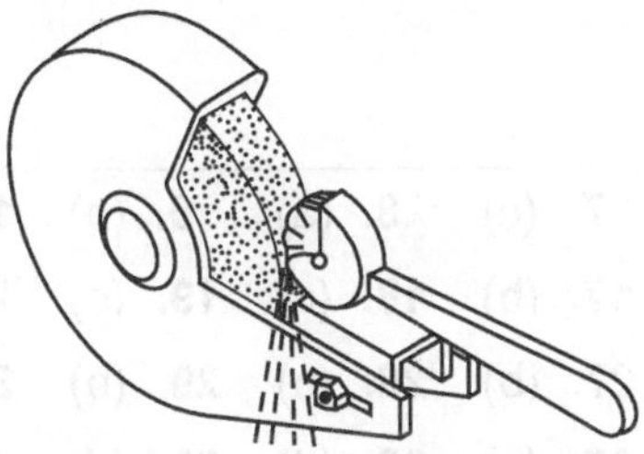

(a) Glazing (b) Loading
(c) Dressing (d) Pinning

83. What is the recommended gap between tool post and grinding wheel?
(a) 1 mm (b) 2 mm
(c) 3 mm (d) 4 mm

84. What is the pitch of wave set in hexablade?
(a) 0.4 mm (b) 0.8 mm
(c) 1.2 mm (d) Over 1.0 mm

85. What tools are used to remove drills and sockets from the machine spindle?
(a) Screwdriver and hammer
(b) Spanner and hammer
(c) Punch and hammer
(d) Drift and hammer

86. Which tap in a set has 20° chamfer at its end?
(a) Intermediate tap (b) Taper tap
(c) Plug tap (d) Second tap

87. Why alternate teeth are staggered in hacksaw blade?
(a) For free cutting action
(b) For curve cutting action
(c) For coarse cutting action
(d) For straight cutting action

88. The point angle for twist drill for general purpose is:
(a) 135° (b) 118°
(c) 90° (d) 60°

89. Twist drills are made of:
(a) Mild steel
(b) High carbon steel
(c) Low carbon steel
(d) Stainless steel

90. What is the unit of cutting speed?
(a) Metre per minute
(b) Metre per second
(c) Centimetre per minute
(d) Centimetre per second

91. Which type of abrasive removes heavy stock of material?
(a) Emery sheet
(b) Boron carbide
(c) Aluminium oxide
(d) Silicon carbide

92. Which type of vice is used to hold a job for drilling?
(a) Bench vice (b) Leg vice
(c) Machine vice (d) Hand vice

93. Which part of a drill forms the cutting edge?
(a) Land (b) Point
(c) Flute (d) Body

94. Which part of drill separates the flutes?
(a) Lip (b) Land
(c) Cutting edge (d) Web

95. Which angle of drill bit prevents friction behind the cutting edge?
(a) Point angle (b) Rake angle
(c) Clearance angle (d) Helix angle

96. Which tool is used to make internal threads in a component?
(a) Reamer (b) Drill
(c) Die (d) Tap

97. Which tool is used to make external threads on a pipe?
(a) Tap (b) Drill
(c) Die (d) Reamer

98. Which is used to hold a die?
(a) Wrench (b) Spanner
(c) Stock (d) Chuck

99. Which of these is a multi-point cutting tool?
(a) Chisel (b) Scraper
(c) Reamer (d) Snips

100. What is probable cause if a twist drill does not cut?
(a) Clearance angle is too small
(b) Clearance angle is too big
(c) Rake angle is too small
(d) Rake angle is too big

ANSWERS

1. (b)	2. (b)	3. (d)	4. (b)	5. (a)	6. (b)	7. (c)	8. (a)	9. (b)	10. (b)
11. (a)	12. (a)	13. (c)	14. (a)	15. (b)	16. (a)	17. (b)	18. (c)	19. (c)	20. (a)
21. (c)	22. (c)	23. (c)	24. (d)	25. (d)	26. (c)	27. (b)	28. (a)	29. (b)	30. (d)
31. (d)	32. (d)	33. (c)	34. (a)	35. (b)	36. (c)	37. (b)	38. (d)	39. (d)	40. (b)
41. (b)	42. (a)	43. (b)	44. (c)	45. (a)	46. (d)	47. (d)	48. (a)	49. (b)	50. (a)
51. (b)	52. (b)	53. (a)	54. (d)	55. (d)	56. (c)	57. (b)	58. (d)	59. (c)	60. (c)
61. (b)	62. (b)	63. (b)	64. (b)	65. (c)	66. (d)	67. (b)	68. (c)	69. (a)	70. (a)
71. (a)	72. (d)	73. (b)	74. (a)	75. (a)	76. (c)	77. (a)	78. (a)	79. (d)	80. (d)
81. (c)	82. (c)	83. (b)	84. (b)	85. (d)	86. (c)	87. (a)	88. (b)	89. (b)	90. (a)
91. (d)	92. (c)	93. (c)	94. (d)	95. (c)	96. (d)	97. (c)	98. (c)	99. (c)	100. (a)

EXPLANATORY NOTES

2: **Tab Washers:** These are thin metal washers with two or more tabs or projections, one of which is bent over the job, or fits into a hole in the job, the other being bent up against a face of the nut. It is not permissible to straighten the tabs and use the washer again: each washer must be used once only.

3: A countersunk screw is a type of fastening that sits flush with the surface of the material it occupies.

11: See other three types of head screws in the figures given below.

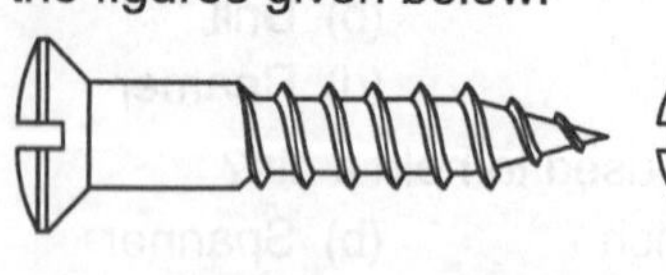

Slotted raised countersunk head screw

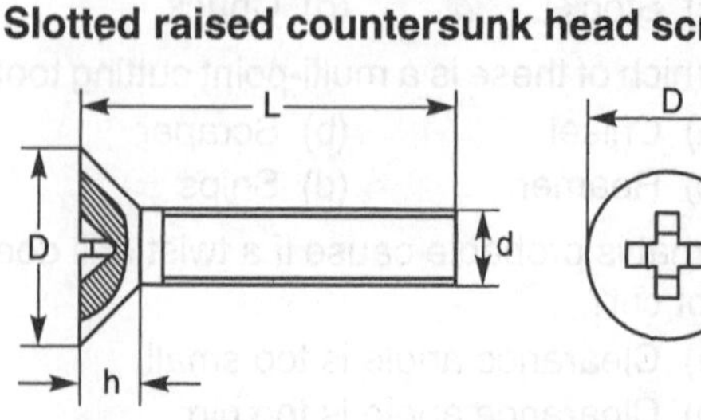

Cross recessed countersunk head screw

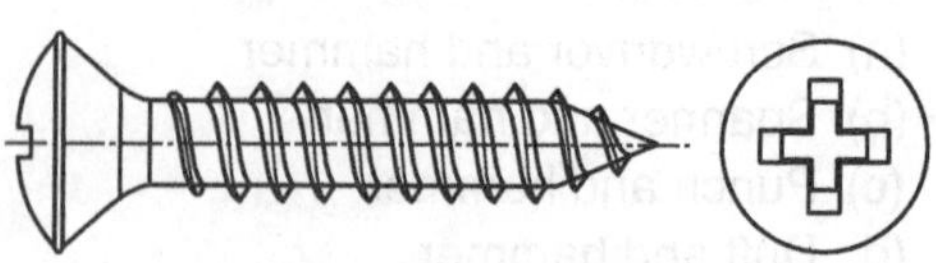

Cross recessed raised countersunk head screw

30: Internal and external circlips are shown in figure given below.

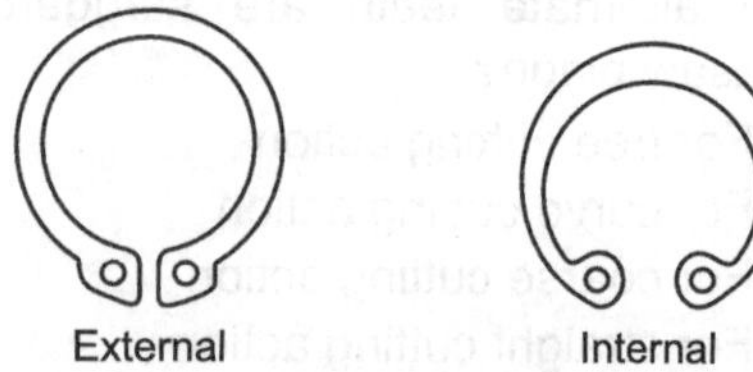

Internal and external ciclips

32: The body-bound bolt, also called the tight-fitting bolt, is one which is turned in a lathe to fit closely a bored hole. Bolts of this kind are much used in holding together parts of machinery to prevent the slightest change of position of one part on another.

33: Gib-head key is tapered on its upper surface and is driven in to form a very secure fastening. The shape of the head provides easy removal.

41: Abrasive machining is a machining process where material is removed from a workpiece using a multitude of small abrasive particles. Common examples include grinding, honing, and polishing.

Laser fusion cutting is used when parts have to meet high optical requirements without finishing. In fusion cutting, the laser beam melts the metal along the whole cutting contour while a high pressure inert gas stream blows the molten material out of the kerf.

51: The safe edge of a file is shown in figure given below.

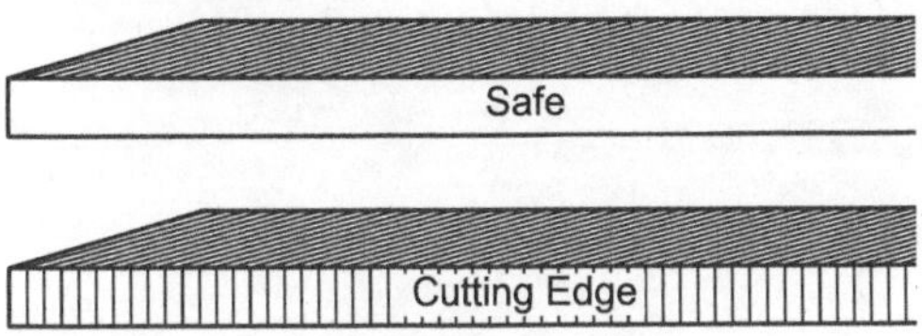

Safe edge of a file

52: To prevent the hacksaw blade binding when penetrating into the material, and to allow free movement of the blade, the cut is to be broader than the thickness of the hacksaw blade. This is achieved by the setting of the hacksaw teeth.

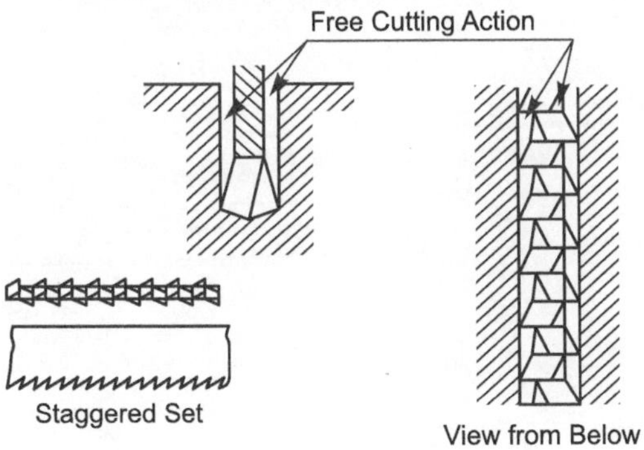

Staggered set

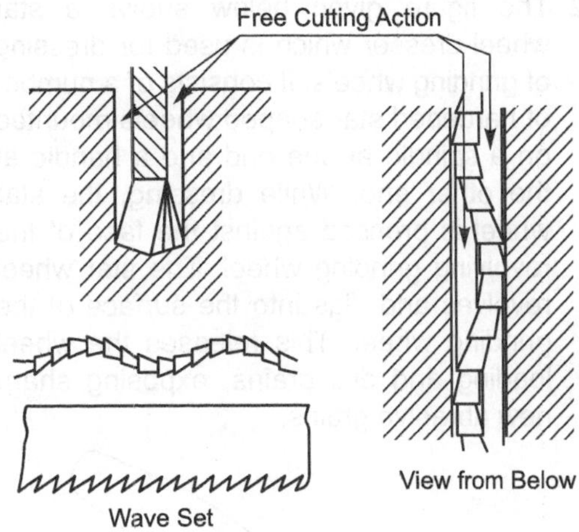

Wave set

59: Adjustable frame hacksaw is the most commonly used type. It gives a better grip and control, while sawing.

62: Breast Drill Machine is a portable drill with a plate that is pressed against the chest to force the drill point into the work. These drill machines are suitable for small drilling applications. Can be used to drill in wood, plastic, mild steel depending on the drill bit used.

A pneumatic drilling machine is designed to perform functions other than drilling, such as reaming, countersinking, and tapping small or large holes.

64: The parts of a drill bit are shown in the figure given below.

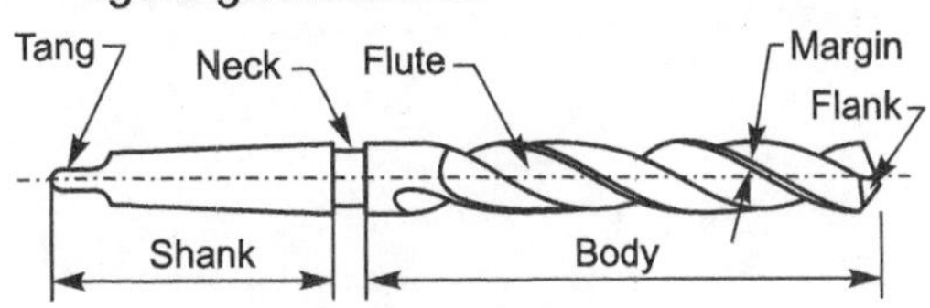

Parts of drill bit

82 The figure given below shows a star wheel dresser which is used for dressing of grinding wheels. It consists of a number of hardened star-shaped wheels mounted on a spindle at one end and a handle at the other end. While dressing, the star wheel is pressed against the face of the revolving grinding wheel. The star wheel revolves and digs into the surface of the grinding wheel. This releases the wheel loading and dull grains, exposing sharp new abrasive grains.

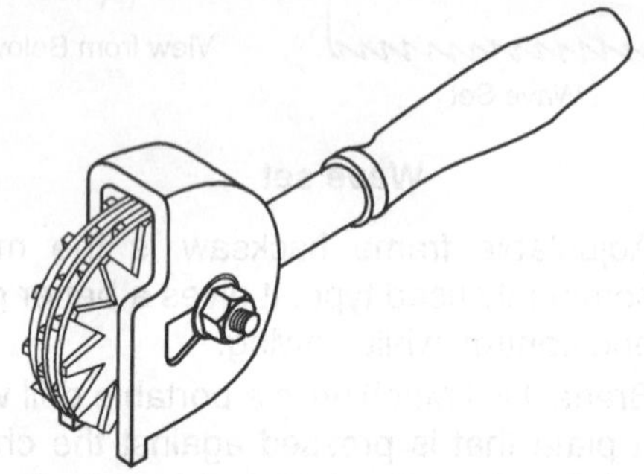

Star wheel dresser

85 The figure given below shows how a drift is used to remove drill from machine spindle.

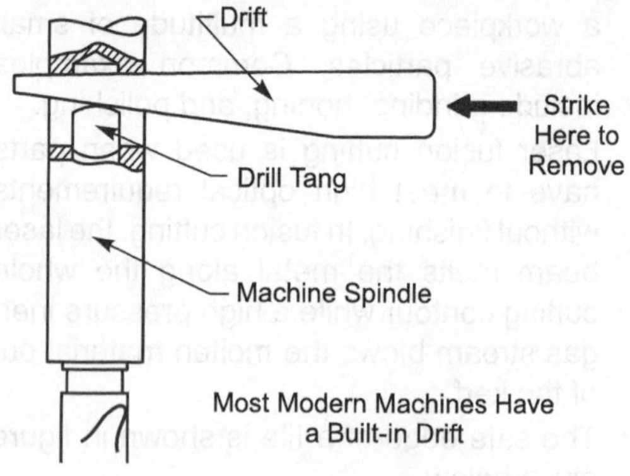

Using drill drift

Module 4

Electrical and Electronics

MULTIPLE CHOICE QUESTIONS

1. An ammeter measures:

(a) Power (b) Voltage
(c) Current (d) Resistance

2. What is the unit of capacitance?

(a) Ohm (b) Voltage
(c) Farad (d) Ampere

3. What is the name of the instrument shown in figure given below?

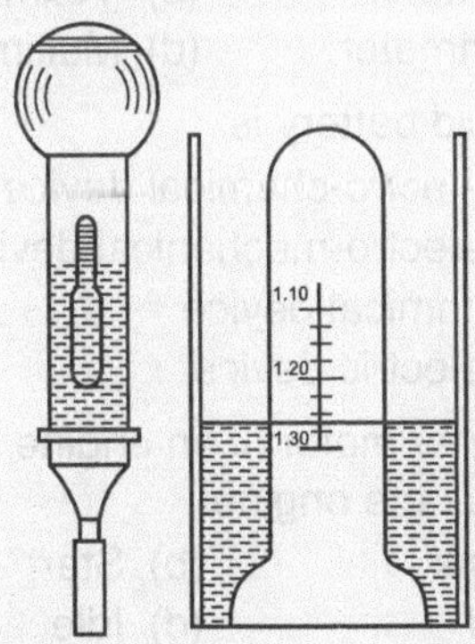

(a) Hydrometer (b) Thermometer
(c) Lactometer (d) Barometer

4. What is the logic gate shown in figure given below?

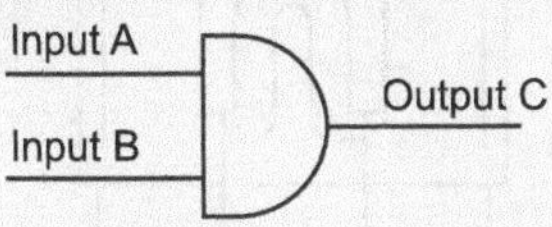

(a) AND gate (b) OR gate
(c) NOT gate (d) NAND gate

5. Name the gate shown in figure given below.

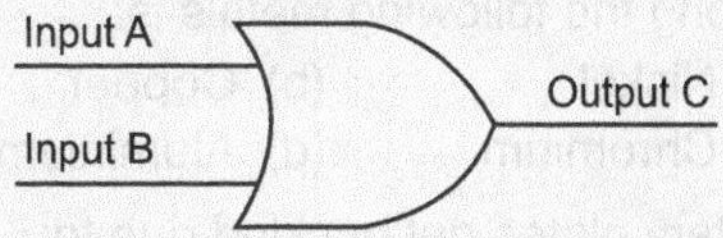

(a) AND gate (b) OR gate
(c) NOT gate (d) NAND gate

6. Name the symbol shown in figure given below.

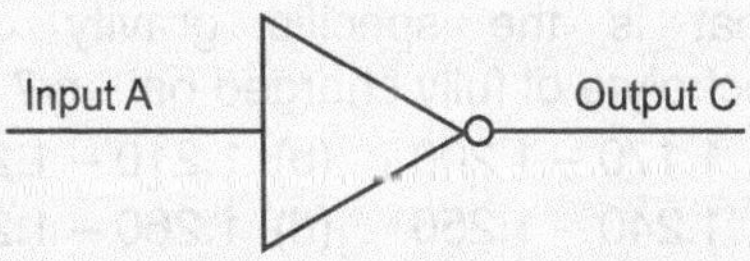

(a) OR gate (b) AND gate
(c) NOT gate (d) NAND gate

7. Which logic gate is called as "Inverter"?

(a) OR gate (b) AND gate
(c) NOT gate (d) NOR gate

8. Which material is required to make integrated circuit chips?

(a) Zinc (b) Steel
(c) Silicon (d) Copper

9. What is the name of the electronic symbol shown in figure given below?

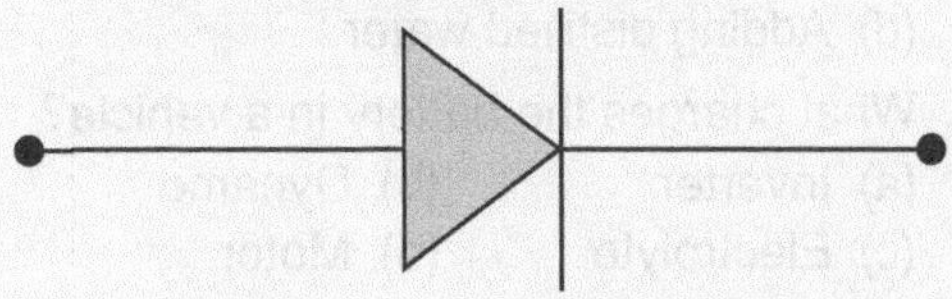

(a) Diode (b) Switch
(c) Battery (d) Transistor

10. Which electronic component is used as a solid state switch?
(a) Inductor (b) Resistor
(c) Capacitor (d) Transistor

11. A transistor has:
(a) One PN junction
(b) Two PN junctions
(c) Three PN junctions
(d) Four PN junctions

12. What type of logic gate is a shuttle valve?
(a) OR gate (b) AND gate
(c) NOR gate (d) NAND gate

13. The best conductor of heat and electricity among the following metals is:
(a) Nickel (b) Copper
(c) Chromium (d) Aluminium

14. Battery plates get buckled due to:
(a) Low electrolyte specific gravity
(b) Over-filled electrolyte
(c) Overcharging
(d) Higher electrolyte specific gravity

15. What is the specific gravity of the electrolyte of fully charged battery?
(a) 1.170 – 1.200 (b) 1.210 – 1.230
(c) 1.240 – 1.250 (d) 1.260 – 1.280

16. What for is the electrical solenoid provided on some fuel injection pumps?
(a) To heat the incoming fuel
(b) To provide excess fuel for starting
(c) To stop the engine through ignition switch and cut off fuel
(d) For electrical governing

17. During battery discharge, the electrolyte is compensated by:
(a) Lead peroxide
(b) Lead sulphate
(c) Oxygen
(d) Adding distilled water

18. What charges the battery in a vehicle?
(a) Inverter (b) Dynamo
(c) Electrolyte (d) Motor

19. SCR stands for:
(a) Selective Catalytic Reduction
(b) Selective Catalytic Regulator
(c) Silicon Catalytic Regulator
(d) Silicon Catalytic Recirculation

20. Which component can convert AC produced by an alternator into DC?
(a) Triode (b) Capacitor
(c) Diode (d) Insulator

21. Where the starter motor is fixed in an engine?
(a) Front side of the engine
(b) Near intake manifold
(c) Near exhaust manifold
(d) Rear side of the engine

22. What type of winding is best suitable to produce a high starting torque in starter motor?
(a) Series winding
(b) Shunt winding
(c) Compound winding
(d) Parallel winding

23. Which of the following instruments can be used for measuring resistance?
(a) Ammeter (b) Voltmeter
(c) Wattmeter (d) Multimeter

24. Lead acid battery is:
(a) An electro-chemical device
(b) An electro-mechanical device
(c) A chemical device
(d) An electric device

25. The starter motor of an engine is meant to .... the engine.
(a) Crank (b) Start
(c) Stall (d) Idle

26. What is the name of the electrical part shown in figure given below?

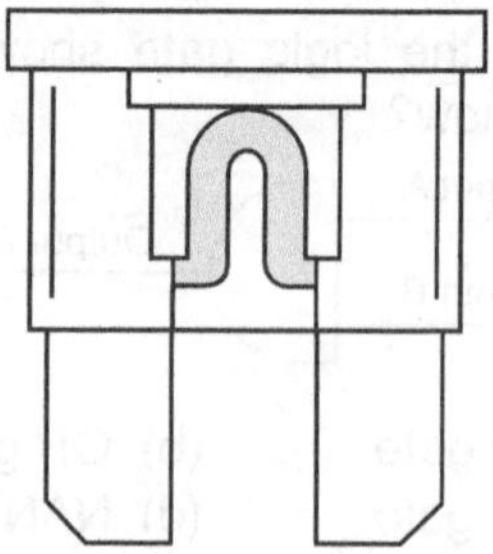

(a) Fuse (b) Plug
(c) Bulb (d) Relay

27. Which device has the ability to store electrical charge?
(a) Capacitor (b) Resistor
(c) Insulator (d) Conductor

28. What is the energy conversion of battery during discharge?
(a) Electrical energy into heat energy
(b) Chemical energy into electrical energy
(c) Electrical energy into chemical energy
(d) Electrical energy into mechanical energy

29. What is the material of positive plate in the lead acid battery?
(a) Tin (b) Lead peroxide
(c) Antimony (d) Spongy lead

30. Which acid is used in the lead acid battery?
(a) Nitric acid
(b) Sulphuric acid
(c) Hydrobromic acid
(d) Hydrochloric acid

31. What is the name of the symbol shown in figure given below?

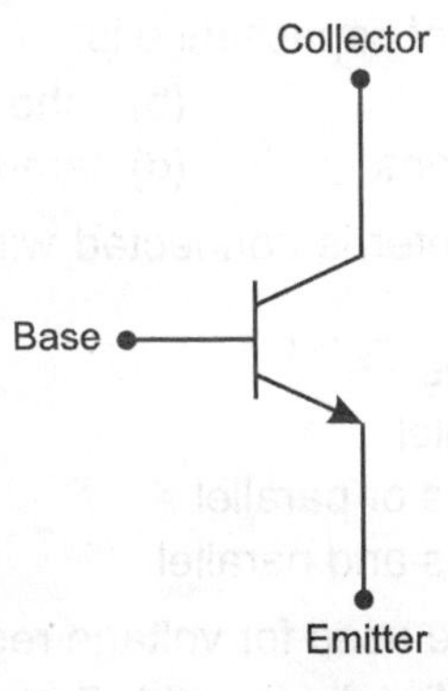

(a) Diode (b) Transistor
(c) Triode (d) Thermistor

32. How many cells are there in a 12 V battery?
(a) 2 cells (b) 6 cells
(c) 9 cells (d) 12 cells

33. What is the colour of positive plates in the charged lead - acid - battery?
(a) Grey (b) Brown
(c) White (d) Black

34. How is the capacity of a battery expressed?
(a) Ampere-hour rating
(b) Voltage-hour rating
(c) Ampere rating
(d) Voltage rating

35. What is the name of the circuit shown in figure given below?

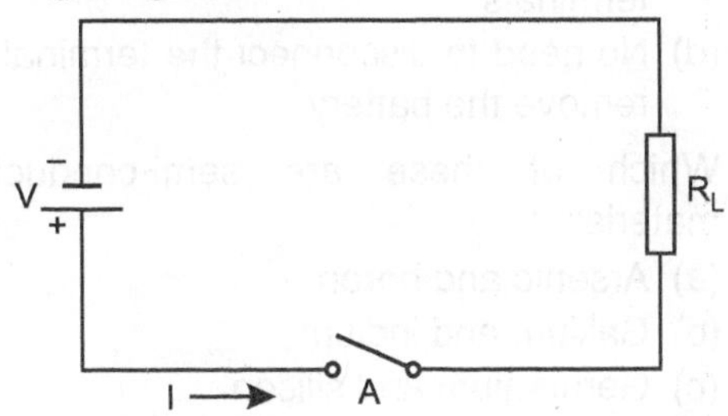

(a) Open circuit (b) Short circuit
(c) Closed circuit (d) Parallel circuit

36. Which law states that "Current is directly proportional to voltage and inversely proportional to resistance"?
(a) Ohm's Law (b) Hooke's Law
(c) Boyle's Law (d) Newton's Law

37. What is the ampere hour rating of a battery that delivers 5 ampere for 20 hours?
(a) 80 ampere hour
(b) 90 ampere hour
(c) 100 ampere hour
(d) 110 ampere hour

38. Which of these components is made of semi-conductor material?
(a) Switch (b) Resistor
(c) Capacitor (d) Transistor

39. What is the name of the central part of an atom?
(a) Proton (b) Electron
(c) Neutron (d) Nucleus

40. Which electrical measuring instruments is fitted on the vehicle panel board?
(a) Ammeter (b) Voltmeter
(c) Wattmeter (d) Ohm-meter

41. In which circuit a ballast resistance is used?
(a) Horn circuit
(b) Wiper circuit
(c) Flasher circuit
(d) Head lamp circuit

42. What is the advantage of maintenance-free sealed battery?
(a) No need to check and top up distilled water
(b) No need for recharging of battery
(c) No chance of sulphation of battery terminals
(d) No need to disconnect the terminal to remove the battery

43. Which of these are semi-conductor materials?
(a) Arsenic and boron
(b) Gallium and indium
(c) Germanium and silicon
(d) Aluminium and antimony

44. Which device is used to step up the voltage in the spark ignition system?
(a) Ignition coil
(b) Spark plug
(c) Condenser
(d) High tension cable

45. What battery charge rate is indicated by the float of the hydrometer (figure given below)?

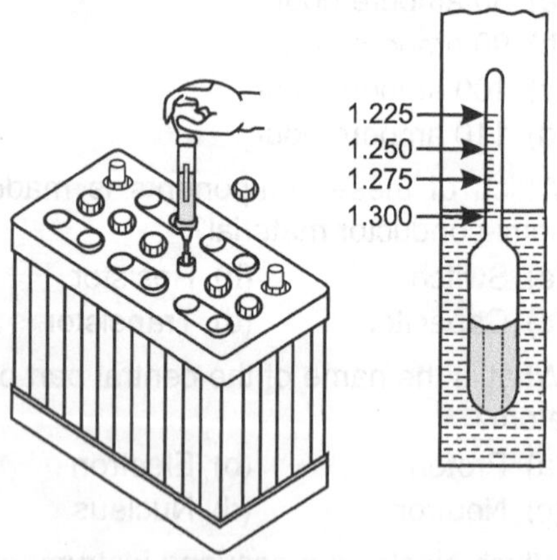

(a) Discharged (b) ¼ charge
(c) ½ charge (d) Full charge

46. How the number of positive and negative plates are compared in a lead acid Battery?
(a) Number of positive and negative plates is equal
(b) Number of negative plates is one more than that of positive plates
(c) Number of negative plates is one less than that of positive plates
(d) None of the above

47. What is an ohmmeter used for in an electrical circuit?
(a) Measure resistance
(b) Measure current
(c) Measure voltage
(d) Measure power

48. Which instrument is used to check the continuity of the field coils of a DC motor?
(a) Ammeter (b) Voltmeter
(c) Ohmmeter (d) Wattmeter

49. Which circuit has an infinite resistance?
(a) Open circuit
(b) Short circuit
(c) Series closed circuit
(d) Parallel closed circuit

50. Which circuit can lead to a hazardous effect?
(a) Open circuit (b) Series circuit
(c) Short circuit (d) Parallel circuit

51. The unit of capacitance is:
(a) ohm (b) mho
(c) siemens (d) farad

52. An ammeter is connected with the circuit in:
(a) Series
(b) Parallel
(c) Series or parallel
(d) Series and parallel

53. The diode used for voltage regulation is:
(a) Rectifier diode (b) Zener diode
(c) Signal diode (d) Varicap diode

54. NPN transistors are preferred to PNP transistors, because of:
(a) Higher switching speed
(b) Wide range of operating temperatures
(c) Easy use in +ve supply rail
(d) (a) and (b) above

55. The process of introducing some types of impurity in germanium and silicon crystals is called:
(a) Dopple (b) Doping
(c) Dumping (d) Dosage

ANSWERS

1. (c)	**2.** (c)	**3.** (a)	**4.** (a)	**5.** (b)	**6.** (c)	**7.** (c)	**8.** (c)	**9.** (a)	**10.** (d)
11. (b)	**12.** (a)	**13.** (b)	**14.** (c)	**15.** (d)	**16.** (c)	**17.** (d)	**18.** (b)	**19.** (a)	**20.** (c)
21. (d)	**22.** (a)	**23.** (d)	**24.** (a)	**25.** (a)	**26.** (a)	**27.** (a)	**28.** (b)	**29.** (b)	**30.** (b)
31. (b)	**32.** (b)	**33.** (b)	**34.** (a)	**35.** (a)	**36.** (a)	**37.** (c)	**38.** (d)	**39.** (d)	**40.** (a)
41. (c)	**42.** (a)	**43.** (c)	**44.** (a)	**45.** (d)	**46.** (b)	**47.** (a)	**48.** (c)	**49.** (a)	**50.** (c)
51. (d)	**52.** (a)	**53.** (b)	**54.** (a)	**55.** (b)					

EXPLANATORY NOTES

3: Hydrometer is used to measure relative density of liquids. Thermometer is used to measure temperature. Lactometer is an instrument used to measure density of milk. Barometer is used to measure atmospheric pressure.

19: **Selective Catalytic Reduction (SCR)** is an advanced active emissions control technology system that injects a liquid-reductant agent through a special catalyst into the exhaust stream of a diesel engine. It is called "selective" because it reduces levels of NOx using ammonia as a reductant within a catalyst system.

23: A *multimeter* is an electronic measuring instrument that combines several measurement functions in one unit. A typical *multimeter* can measure voltage, current, and resistance.

24: An electro-chemical device is capable of either generating electrical energy from chemical reactions or using electrical energy to cause chemical reactions.

28: Note that during charging of the battery, electrical energy is converted into chemical energy.

32: Each cell produces approximately 2.1 volts.

41: A ballast resistor is a resistor that has the property of increasing in resistance as current flowing through it increases, and decreasing in resistance as current decreases.

46: All batteries have one more negative plate than positive plate. That is because the positive plate is the working plate and if there isn't a negative plate on the outside of the last positive plate, the whole outer side of last positive plate will not have anything with which to react and create electricity.

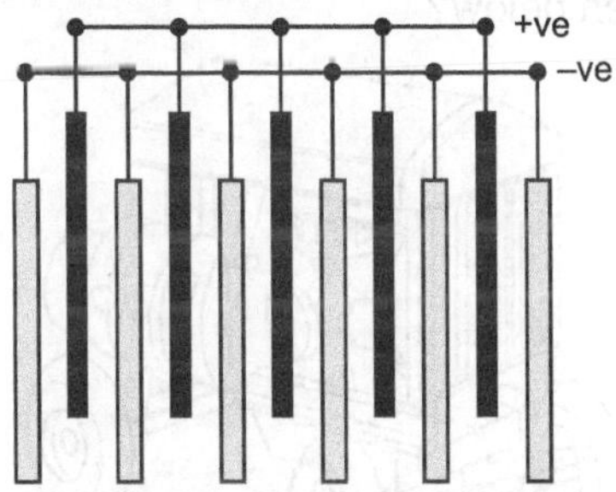

Arrangement of plates in a lead-acid battery

54: One of the major difference between the NPN and PNP transistor is that in the NPN transistor the current flow between collector to emitter when the positive supply is given to the base, whereas in PNP transistor the charge carrier flows from the emitter to collector when negative supply is given to the base.

55: Doping is the process of adding impurities to intrinsic semiconductors to alter their properties. Normally Trivalent and Pentavalent elements are used to dope Silicon and Germanium.

Module 5

Arc and Gas Welding

MULTIPLE CHOICE QUESTIONS

1. What quenching medium is used for normalising?

(a) Water spray (b) Open air
(c) Oil bath (d) Water bath

2. What welding device is shown in figure given below?

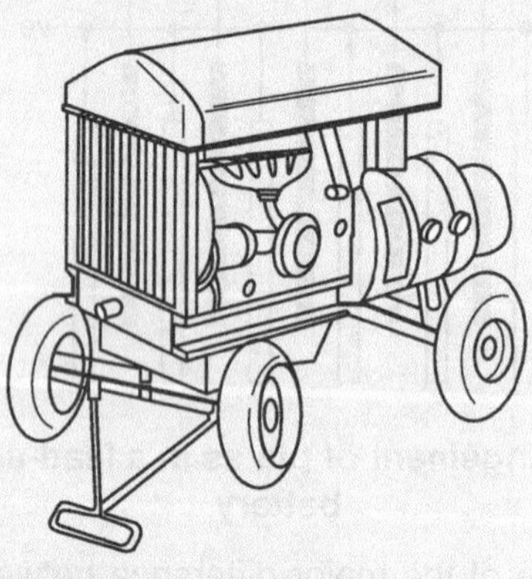

(a) Motor generator
(b) Engine generator
(c) Regulator
(d) Alternator

3. What edge preparation for welding of plates is shown in the figure give below?

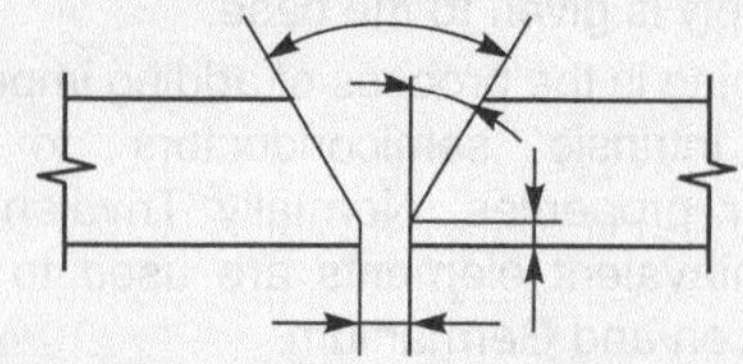

(a) Single 'V' (b) Double 'V'
(c) Single 'U' (d) Double 'U'

4. What physical transfer takes place during arc welding process?

(a) Colour transfer (b) Metal transfer
(c) Current transfer (d) Heat transfer

5. What is the colour of acetylene gas cylinder?

(a) Red (b) Maroon
(c) Black (d) Peacock blue

6. What is the colour of oxygen gas cylinder?

(a) Red (b) Maroon
(c) Black (d) Peacock blue

7. What is the name of the oxy-acetylene flame shown below?

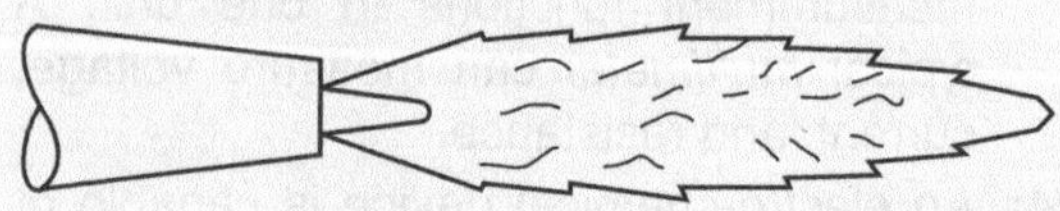

(a) Arc flame
(b) Neutral flame
(c) Carburising flame
(d) Oxidising flame

8. What type of weld joint is shown in the figure given below?

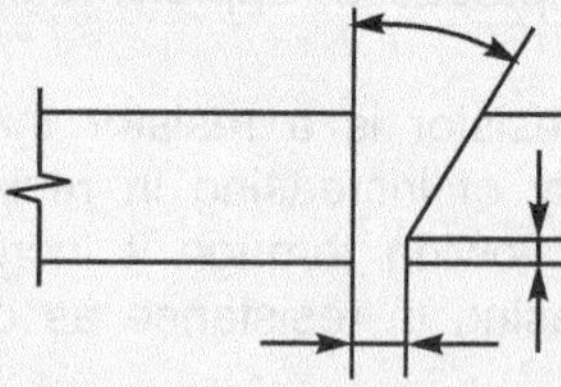

(a) Double 'V' butt (b) Double 'U' butt
(c) Single bevel butt (d) Double 'J' butt

9. Soft solder contains lead and:
(a) Silver (b) Tin
(c) Copper (d) Aluminium

10. The welding machine which converts AC supply to DC welding supply is:
(a) Motor generator set
(b) Engine generator set
(c) Rectifier set
(d) Welding transformer

11. What heat treatment process a crankshaft is subjected to?
(a) Carburizing (b) Tempering
(c) Annealing (d) Nitriding

12. The process of inducing toughness and reducing brittleness by heat treatment is called ....
(a) Annealing (b) Hardening
(c) Normalising (d) Tempering

13. In which part of an oxy-acetylene flame is the temperature the highest?
(a) At tho inner cone
(b) At the outer cone
(c) At the tip of the torch
(d) None of the above

14. Name the wolding process as shown in the figure given below.

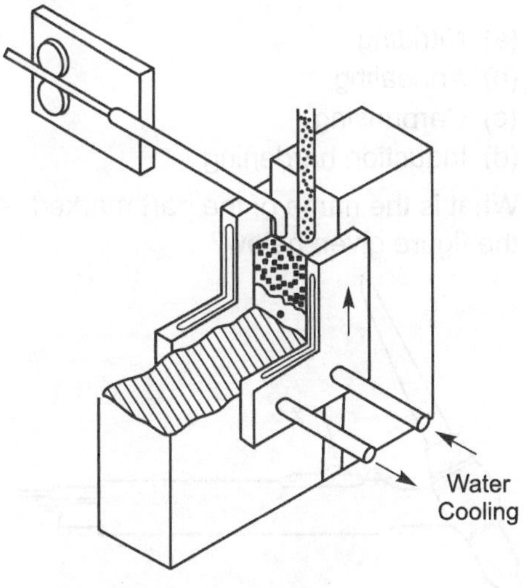

(a) Plasma arc welding
(b) Electroslag welding
(c) Gas metal arc welding
(d) Submerged arc welding

15. What is the diameter of medium coated MS electrode used in fillet weld lap joint with 110/120 Amps welding current?
(a) 1.0 mm (b) 1.5 mm
(c) 2.5 mm (d) 3.15 mm

16. Identify the welding machine shown in the figure given below.

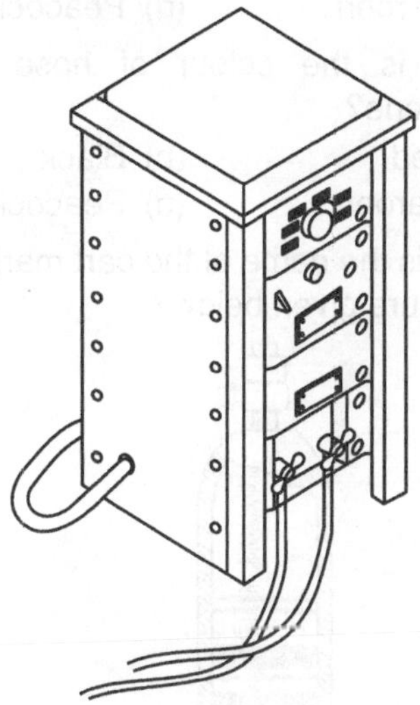

(a) Motor generator set
(b) Engine generator set
(c) DC generator machine
(d) Rectifier set

17. What is the name of the part marked 'X' in the figure given below?

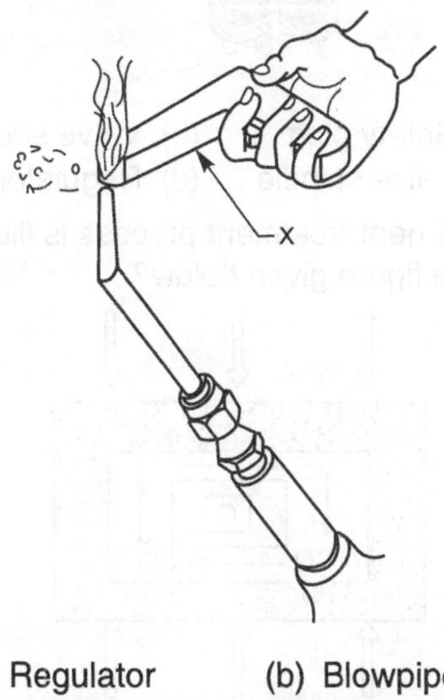

(a) Regulator (b) Blowpipe
(c) Spark lighter (d) Non-return disc

18. What is used for acetylene gas connection leakage test?
(a) Kerosene (b) Naked fire
(c) Soap water (d) Fresh water

19. What is the colour of hose carrying oxygen?
(a) Red (b) Black
(c) Maroon (d) Peacock blue

20. What is the colour of hose carrying acetylene?
(a) Red (b) Black
(c) Maroon (d) Peacock blue

21. What is the name of the part marked 'X' in the figure given below?

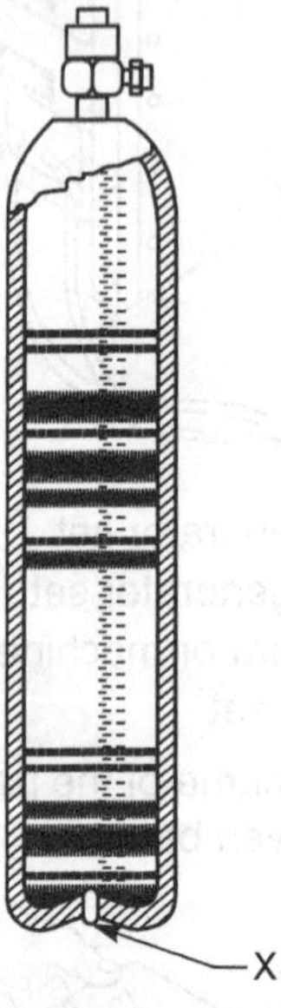

(a) Safety plug (b) Valve socket
(c) Valve spindle (d) Regulator

22. What heat treatment process is illustrated in the figure given below?

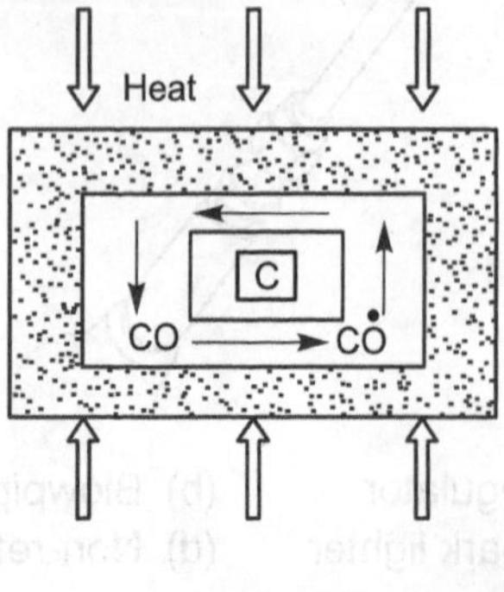

(a) Annealing (b) Carburising
(c) Normalising (d) Tempering

23. Which type of hardening uses eddy current in the metal bar?
(a) Flame hardening
(b) Nitriding hardening
(c) Induction hardening
(d) Carburising hardening

24. Which heat treatment process is used for camshaft production?
(a) Annealing
(b) Carburising
(c) Normalising
(d) Induction hardening

25. What heat treatment process is illustrated in the figure given below?

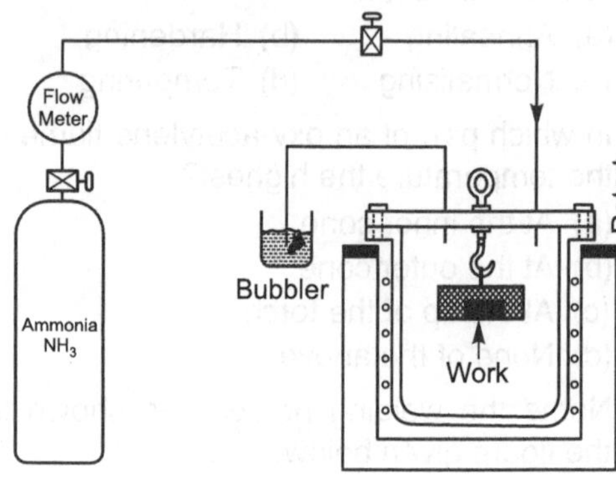

(a) Nitriding
(b) Annealing
(c) Carburising
(d) Induction hardening

26. What is the name of the part marked 'X' in the figure given below?

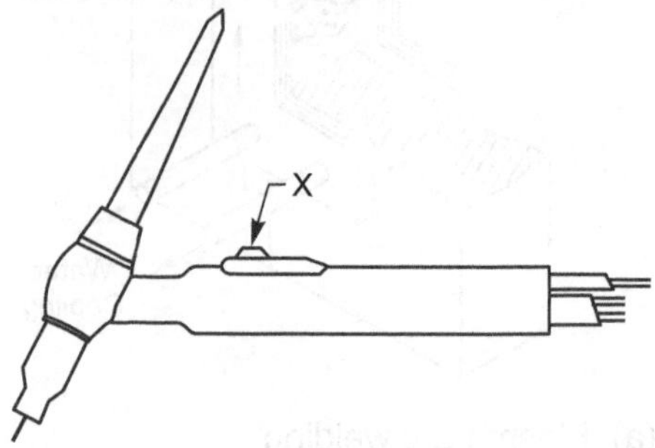

(a) Switch
(b) Ceramic nozzle

(c) Handle
(d) Tungsten electrode

27. What is the name of the part marked 'X' in the figure given below?

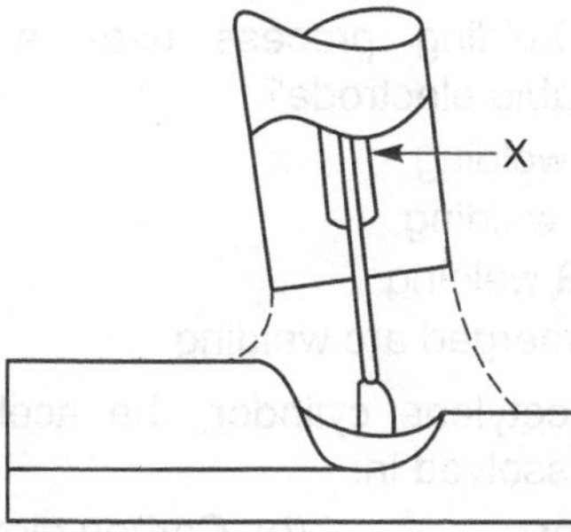

(a) Pilot arc (b) Electrode
(c) Weld pool (d) Contact tube

28. Which of these is a heat treatment process?
(a) Casting (b) Machining
(c) Welding (d) Annealing

29. The arc utilized in arc welding is a:
(a) High voltage, high current discharge
(b) Low voltage, low current discharge
(c) Low voltage, high current discharge
(d) High voltage, low current discharge

30. Which one of the following is a type of transformer used in arc welding?
(a) Step-down
(b) Step-up
(c) One-to-one
(d) Capable of increasing supply voltage

31. In DC welding, heat distribution between electrode and base metal can be changed by changing the polarity. The distribution of heat is:
(a) Positive 2/3 and negative 1/3
(b) Positive 1/3 and negative 2/3
(c) Positive ¾ and negative ¼
(d) Positive ¼ and negative ¾

32. The size of a welding machine is determined by:
(a) Input amperage
(b) Output amperage
(c) Open circuit voltage
(d) Closed circuit voltage

33. The position in which it is easiest to weld is:
(a) Flat (b) Vertical
(c) Horizontal (d) Overhead

34. The size of arc welding electrode refers to:
(a) Diameter of its core wire
(b) Overall diameter of the electrode
(c) Thickness of flux coating
(d) Length of the electrode

35. One of the functions of the coating of an arc welding electrode is to:
(a) Increase welding current
(b) Stabilize the arc
(c) Prevent rusting
(d) Control arc temperature

36. Arc welding electrodes are manufactured in two standard lengths, namely:
(a) 350 mm and 250 mm
(b) 350 mm and 450 mm
(c) 400 mm and 500 mm
(d) 12 inch and 10 inch

37. Which quenching media is applied in case of flame hardening process?
(a) Still air
(b) Oil bath
(c) Brine solution bath
(d) Water spray

38. What is the purpose of hardening?
(a) To increase ductility
(b) To increase malleability
(c) To increase strength
(d) To increase brittleness

39. What is the purpose of annealing?
(a) To improve hardness
(b) To improve machinability
(c) To improve brittleness
(d) To improve weldability

40. What is the advantage of using rectifier set as arc welding machine?
(a) Suitable for welding all ferrous and non-ferrous metals
(b) Low initial cost
(c) Low maintenance cost
(d) No special safety precautions required

41. What is the name of the welding process as shown in the figure given below?

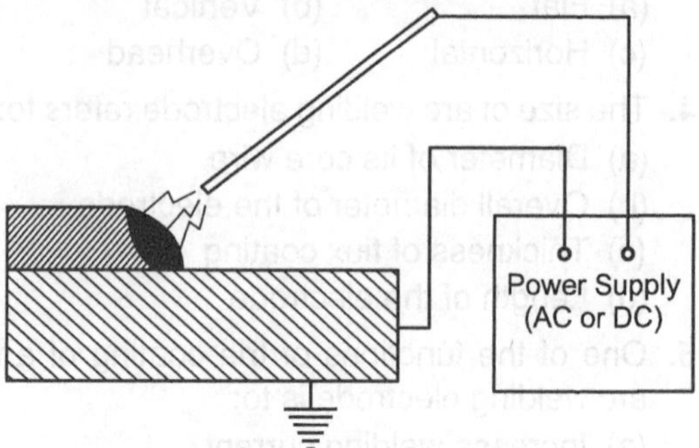

(a) Arc welding
(b) Carbon arc welding
(c) Plasma arc welding
(d) Submerged arc welding

42. Which machine can be used anywhere in the field for welding away from electric main supply?
(a) Motor generator set
(b) Engine generator set
(c) AC arc welding machine
(d) Rectifier set

43. Which one of the following metals cannot be welded using AC welding machine?
(a) Brass
(b) Cast iron
(c) Mild steel
(d) High carbon steel

44. Which of these is safe method for carrying oxyacetylene gas cylinders?
(a) Rolling (b) Sliding
(c) Dragging (d) Using a trolley

45. Which valve should be closed first in case gas flame snaps out and backfires?
(a) Oxygen valve in the cylinder
(b) Oxygen valve at the blowpipe
(c) Acetylene valve in the cylinder
(d) Acetylene valve at the blowpipe

46. Gas tungsten arc welding (GTAW) process is another name for:
(a) TIG welding (b) MIG welding
(c) MAG welding (d) Plasma welding

47. Which of these is a gas metal arc welding (GMAW) process?
(a) TIG welding
(b) MIG welding
(c) MAG welding
(d) MIG and MAG welding both

48. Which welding process uses a non-consumable electrode?
(a) TIG welding
(b) MIG welding
(c) MAG welding
(d) Submerged arc welding

49. In an acetylene cylinder, the acetylene gas is dissolved in:
(a) Water (b) Carbon dioxide
(c) Acetone (d) Mercury

50. Te function of the central hole in the tip of gas cutting blowpipe is to supply:
(a) Oxygen for preheating
(b) Acetylene for preheating
(c) Oxygen for cutting
(d) Acetylene for cutting

51. A carbon steel piece is heated just above 730°C, maintained at that temperature for a few hours and then slowly cooled. What heat treatment process has been done?
(a) Normalising (b) Case hardening
(c) Hardening (d) Annealing

52. To reduce internal stresses of a hardened tool, the method of heat treatment generally applied is:
(a) Stabilising (b) Annealing
(c) Normalising (d) Tempering

53. In a case hardening process, ammonia gas is introduced on steel workpiece; the process is known as:
(a) Cyaniding (b) Nitriding
(c) Carburising (d) Ammonising

54. Cyaniding and nitriding are two methods of:
(a) Hardening (b) Case hardening
(c) Tempering (d) Normalising

55. The instrument used to measure high temperature in the furnace is:
(a) Thermometer (b) Barometer
(c) Calorimeter (d) Pyrometer

56. Which one of the following groups of quenching media is in order of their severity of cooling rate, i.e. from slow to rapid cooling?
(a) Oil, forced air, brine solution
(b) Forced air, oil, brine solution
(c) Brine solution, oil, forced air
(d) Forced air, brine solution, oil

57. Which one of the following processes is adopted to improve the internal structure of steel, which has been subjected to severe hammering?
(a) Hardening (b) Annealing
(c) Normalising (d) Tempering

58. While normalising steel piece should be cooled :
(a) In still air to room temperature
(b) In oil
(c) By forced air
(d) In water

ANSWERS

1. (b)	**2.** (b)	**3.** (a)	**4.** (b)	**5.** (b)	**6.** (c)	**7.** (b)	**8.** (c)	**9.** (b)	**10.** (c)
11. (d)	**12.** (d)	**13.** (a)	**14.** (b)	**15.** (d)	**16.** (d)	**17.** (c)	**18.** (c)	**19.** (b)	**20.** (c)
21. (a)	**22.** (b)	**23.** (c)	**24.** (d)	**25.** (a)	**26.** (a)	**27.** (d)	**28.** (d)	**29.** (c)	**30.** (a)
31. (a)	**32.** (b)	**33.** (a)	**34.** (a)	**35.** (b)	**36.** (b)	**37.** (d)	**38.** (c)	**39.** (b)	**40.** (a)
41. (a)	**42.** (b)	**43.** (a)	**44.** (d)	**45.** (b)	**46.** (a)	**47.** (d)	**48.** (a)	**49.** (c)	**50.** (c)
51. (d)	**52.** (d)	**53.** (b)	**54.** (b)	**55.** (d)	**56.** (b)	**57.** (c)	**58.** (a)		

EXPLANATORY NOTES

1: Normalising is a heat treatment process, of which the aim is to obtain fine-grained, uniformly distributed ferrite – pearlite structure.

14: Other three arc welding processes mentioned in this question are shown in the figures given below.

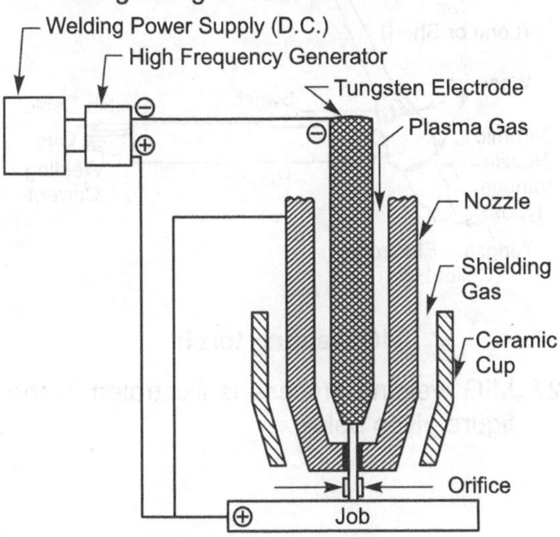

Plasma arc welding

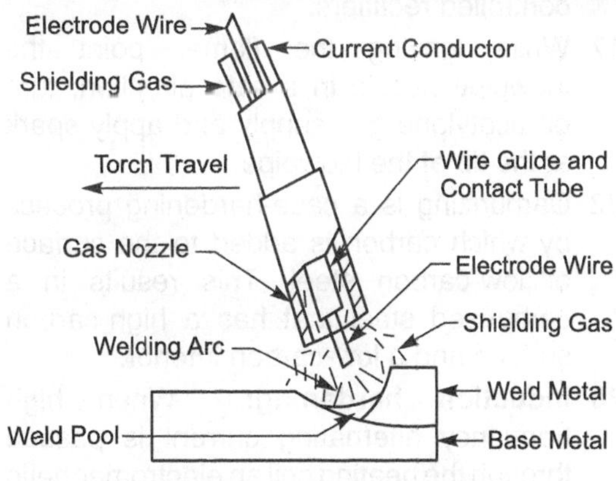

Gas metal arc welding

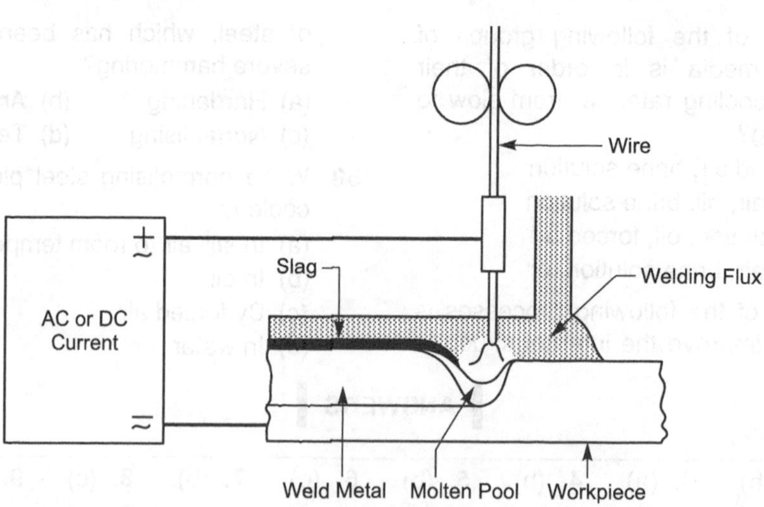

Submerged arc welding

16 Welding Rectifiers produce DC or, AC and DC welding current. They may utilize either single phase or three phase input power. They contain a transformer, but rectify the AC to DC by the use of selenium rectifiers, silicon diodes or silicon controlled rectifiers.

17 While igniting the flame, point the blowpipe nozzle in a safe direction, turn on acetylene gas supply and apply spark at the tip of the blowpipe.

22 Carburizing is a case-hardening process by which carbon is added to the surface of low-carbon steel. This results in a carburized steel that has a high-carbon surface and a low-carbon interior.

23 **Induction hardening:** When high frequency alternating current is passed through the heating coil an electromagnetic field is created around it. It gives rise to eddy currents in the surface of the metal bar centred in the coil. Thus, the surface of the metal bar gets heated above the critical temperature and subsequently gets hardened during quenching.This method is employed for very long parts and normally requires a cross sectional area that is uniform along the entire length of the hardened surface.

25 Nitriding case-hardening method produces the hardest surface of any of the hardening processes. It introduces nitrogen into the surface of steel. Medium carbon steels are generally case hardened by this process. In this, individual parts are heated-treated in a furnace that has an ammonia gas atmosphere. Time of nitriding is long and will be about 70 hours. The case depth is less than 0.5 mm.

26 Different parts of a TIG welding torch are shown in the figure given below.

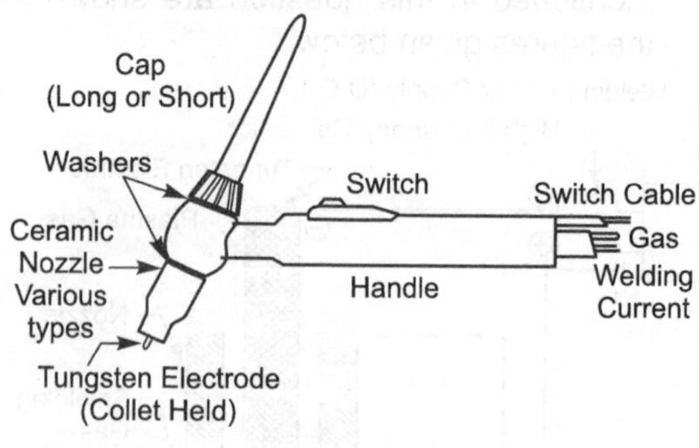

TIG welding torch

27 MIG welding process is illustrated in the figure given below.

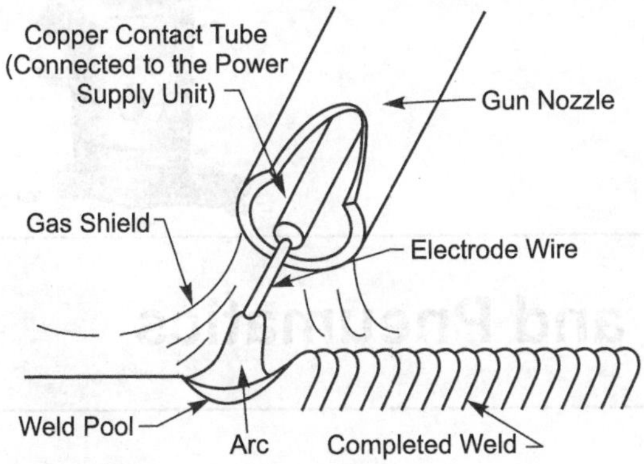

MIG welding

42 Engine generator set is similar to the motor generator set except that the generator is driven by a petrol or diesel engine. Its running and maintenance charges are higher. It can be used anywhere in field work, away from electric lines.

48 In tungsten inert gas arc welding (TIG) the arc is formed between the tungsten electrode (non-consumable) and the welding job in an atmosphere of an inert gas (argon or helium).

50 A gas cutting blowpipe is shown below. The central hole is for oxygen gas for cutting, after the piece has been preheated to the required temperature.

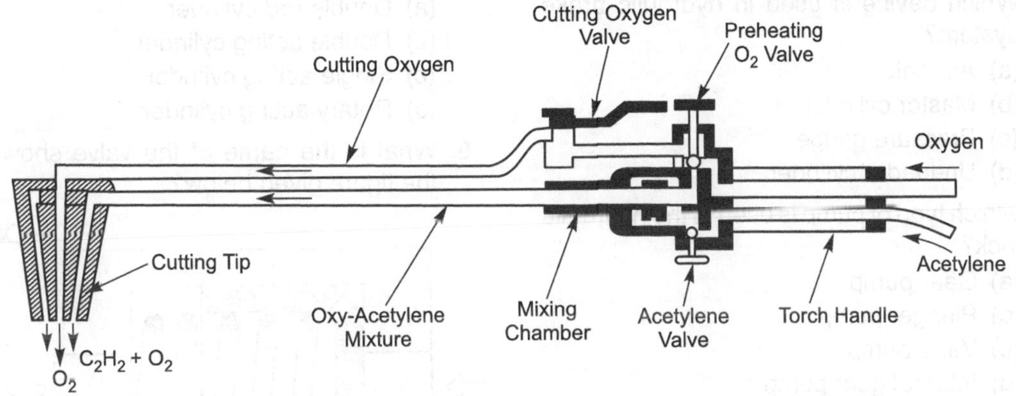

Gas cutting torch

Module 6

Hydraulics and Pneumatics

MULTIPLE CHOICE QUESTIONS

1. Which device is used in hydraulic brake system?
(a) Air tank
(b) Master cylinder
(c) Pressure gauge
(d) Un-loader cylinder

2. Which type of pump is used in the hydraulic jack?
(a) Gear pump
(b) Plunger pump
(c) Vane pump
(d) Internal gear pump

3. Which device converts mechanical into pneumatic energy?
(a) Actuator (b) Generator
(c) Alternator (d) Compressor

4. Which device converts hydraulic energy into mechanical energy?
(a) Actuator (b) Regulator
(c) Compressor (d) Control valve

5. What is the name of the hydraulic cylinder shown in figure given below?

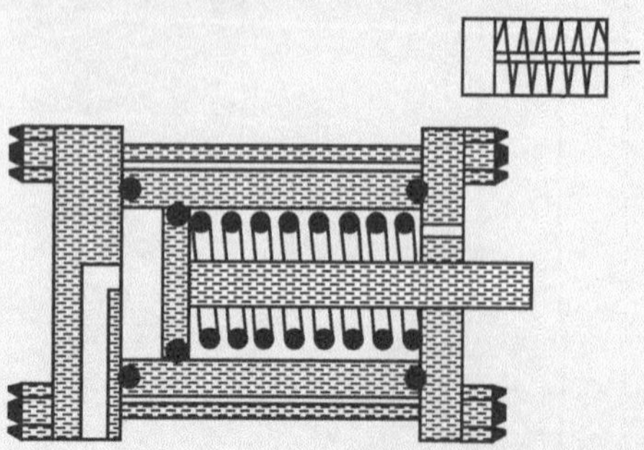

(a) Double rod cylinder
(b) Double acting cylinder
(c) Single acting cylinder
(d) Rotary acting cylinder

6. What is the name of the valve shown in the figure given below?

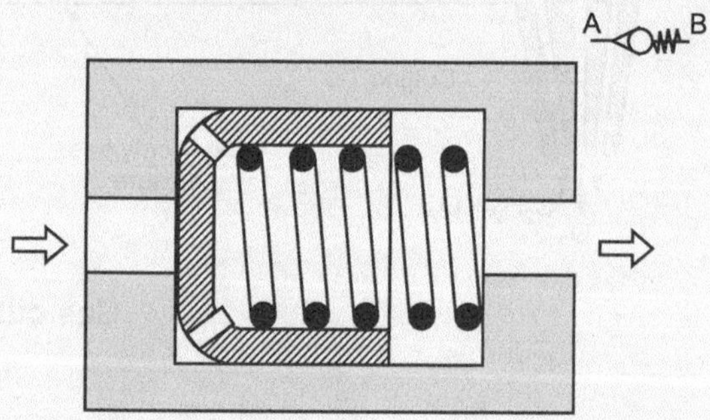

(a) Pressure relief valve
(b) Bypass valve
(c) Check valve
(d) Inlet valve

7. What is the name of the valve shown in the figure given below?

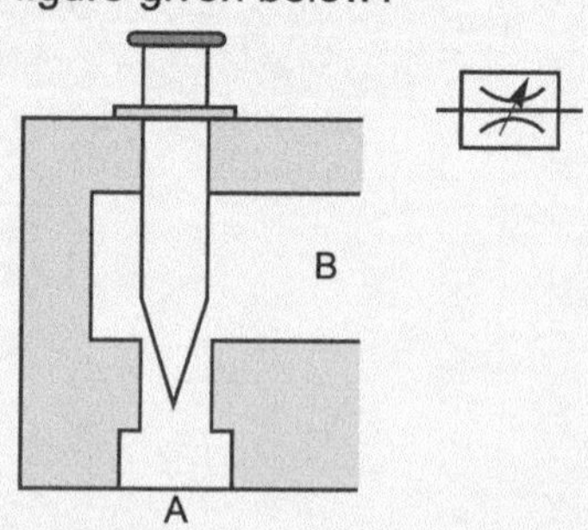

(a) Check valve
(b) Choke valve
(c) Throttle valve
(d) Pressure relief valve

8. What is the type of compressor shown in the figure given below?

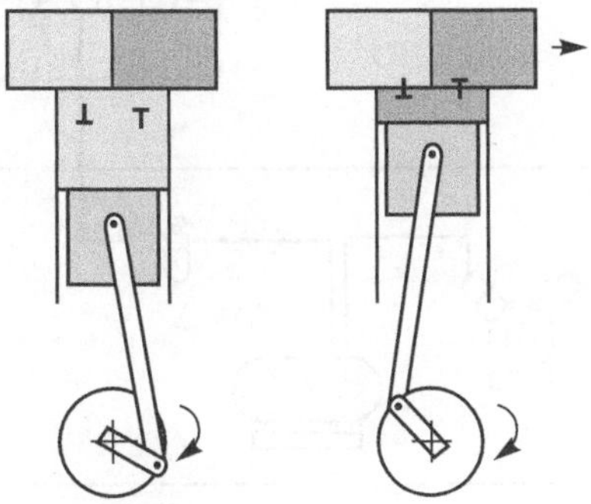

(a) Gear type
(b) Vane type
(c) Rotary type
(d) Reciprocating type

9. What is related to the measurement of resistance to flow of a liquid?

(a) Gravity (b) Elasticity
(c) Quality (d) Viscosity

10. Which law applies to hydraulic brake system?

(a) Boyle's law (b) Charles law
(c) Pascal's law (d) Newton's law

11. What type of pump drive is shown in the figure given below?

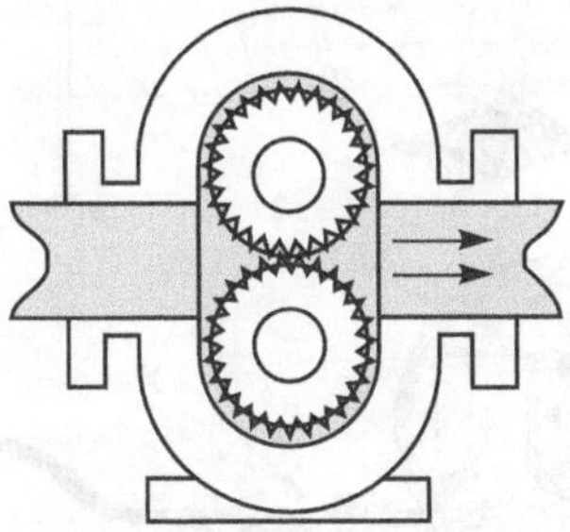

(a) Vane pump
(b) Plunger pump
(c) Internal gear drive pump
(d) External gear drive pump

12. What does the following manual symbol indicate?

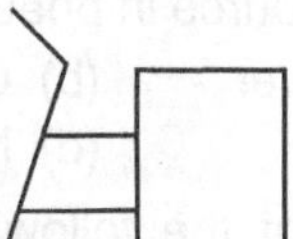

(a) Roller operated (b) Foot pedal
(c) Push button (d) General button

13. What mechanical symbol is shown below?

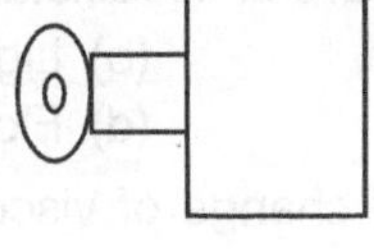

(a) Plunger (b) Push button
(c) Roller (d) Foot pedal

14. What is the name of the part marked 'X' in the figure given below?

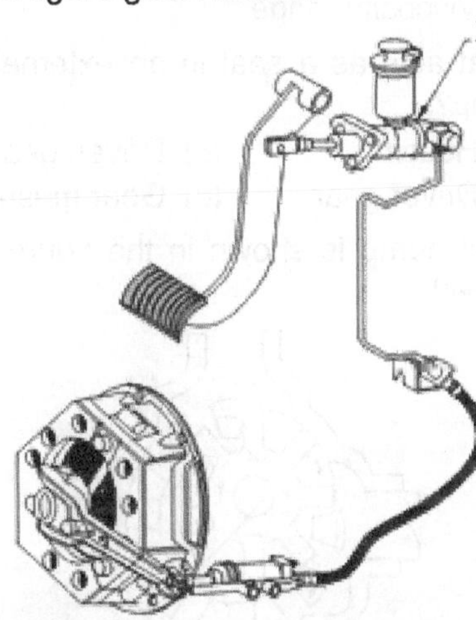

(a) Slave cylinder (b) Master cylinder
(c) Air cylinder (d) Wheel cylinder

15. Which is the power source of a pneumatic system?

(a) Actuator (b) Control valve
(c) Air compressor (d) Air service unit

16. The study of pneumatics deals with systems operated with:

(a) Air (b) Water
(c) Oil (d) Solid

17. In pneumatic system, a sensor serves the purpose of:

(a) Transmission (b) Control
(c) Feedback (d) Storage

18. The power source in pneumatic system is:
(a) Air receiver (b) Compressor
(c) Valve (d) Muffler

19. Which one of the following is a type of actuator in hydraulic system?
(a) Pump (b) Valve
(c) Strainer (d) Cylinder

20. Viscosity is an internal property of a fluid. It is a measure of its resistance to:
(a) Current (b) Light
(c) Sound (d) Flow

21. The rate of change of viscosity of a fluid between two temperatures is called:
(a) Viscosity index
(b) Viscosity coefficient
(c) Viscosity rate
(d) Viscosity range

22. What acts as a seal in an external gear pump?
(a) Housing (b) Driven gear
(c) Driver gear (d) Gear meshing

23. What pump is shown in the figure given below?

(a) Rotor pump
(b) Plunger pump
(c) Internal gear pump
(d) External gear pump

24. Which type of clutch system is used in slave cylinder of brake system?
(a) Vacuum clutch
(b) Hydraulic clutch
(c) Overrunning clutch
(d) Electromagnetic clutch

25. What is the name of the pneumatic component marked 'X' in the figure given below?

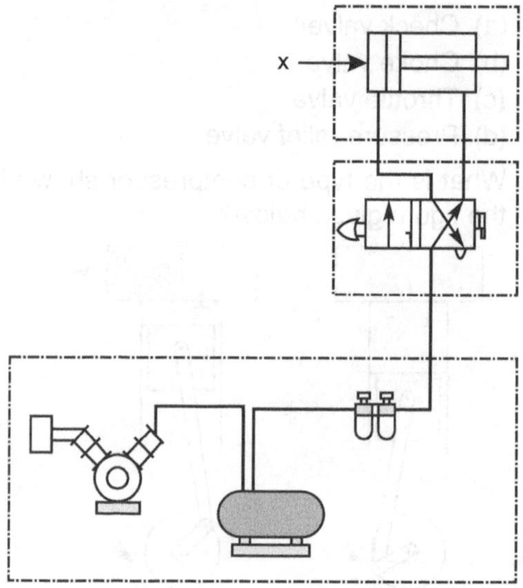

(a) Actuator (b) Compressor
(c) Control valve (d) Reservoir tank

26. Which part in the internal gear pump acts as seal?
(a) Gears
(b) Gasket
(c) Sealing ring
(d) Crescent – shaped spacer

27. What is the name of the part marked 'X' in the hydraulic clutch shown in the figure given below?

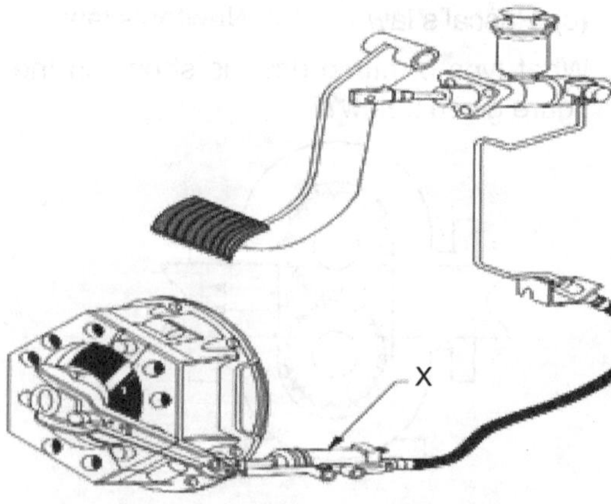

(a) Wheel cylinder (b) Master cylinder
(c) Clutch cylinder (d) Slave cylinder

28. What is the type of compressor shown in figure given below?

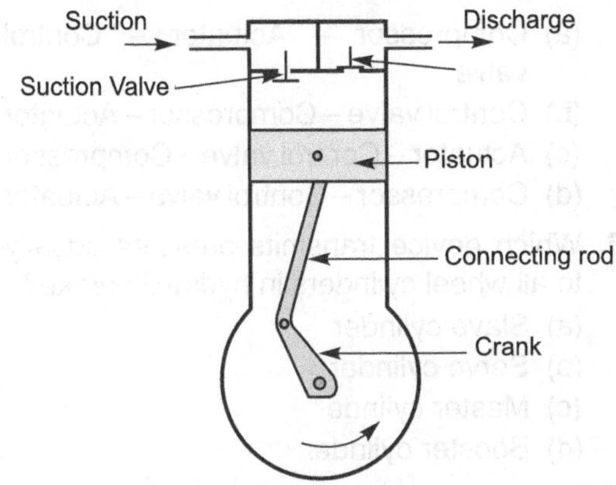

(a) Reciprocating compressor
(b) Turbo compressor
(c) Screw compressor
(d) -Scroll compressor

29. With reference to pneumatic system, what does 'FRL' stand for?
(a) Filler – Reservoir – Lubricator
(b) Filler – Resistor – Lubricator
(c) Filter – Resistor – Lubricator
(d) Filter – Regulator – Lubricator

30. In FRL, what does the following symbol indicate?

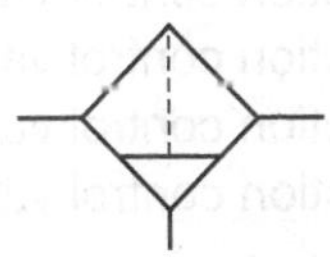

(a) Filter unit (b) Pressureregulator
(c) Lubricator (d) None of these

31. The figure given below shows three actuators. Name them from left to right.

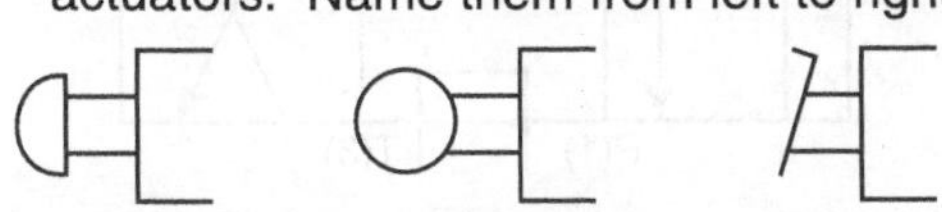

(a) Push button, mechanical, foot pedal
(b) Mechanical, push button, foot pedal
(c) Push button, foot pedal, mechanical
(d) Mechanical, foot pedal, push button

32. 5/2 way single solenoid valve has:
(a) 2 ports, 2 positions
(b) 5 ports, 2 positions
(c) 5 ports, 5 positions
(d) 2 ports, 5 positions

33. A .... restricts air flow.
(a) Throttle valve
(b) Shuttle valve
(c) Directional control valve
(d) Single-acting cylinder

34. The parts of a valve that affect the flow through the valve are collectively called as:
(a) Body (b) Trim
(c) Container (d) Capsule

35. Which valve is used for fine control of flow in pipeline?
(a) Gate valve (b) Globe valve
(c) Needle valve (d) Non-return valve

36. Which valve is used as interface between electrical and pneumatic system?
(a) Roller valve
(b) Flow control valve
(c) Solenoid valve
(d) Pressure relief valve

37. Which valve is used for mechanical position to sense in machine automation system?
(a) Pressure relief valve
(b) Roller valve
(c) Flow control valve
(d) Directional valve

38. Which valve is used to remove the excess amount of oil in hydraulic system?
(a) Pressure relief valve
(b) Pressure reducing valve
(c) Pressure regulator valve
(d) Roller valve

39. Which type of filter is used to remove ferrous materials from oil in hydraulic system?
(a) Pressure line filter
(b) Offline filter
(c) Magnetic filter
(d) Absorbent filter

40. How many switching positions are there in a 4/2 directional control valve?
(a) Four (b) Two
(c) One (d) Three

41. What is the name of the part marked 'X' in the pneumatic system shown in the figure given below?

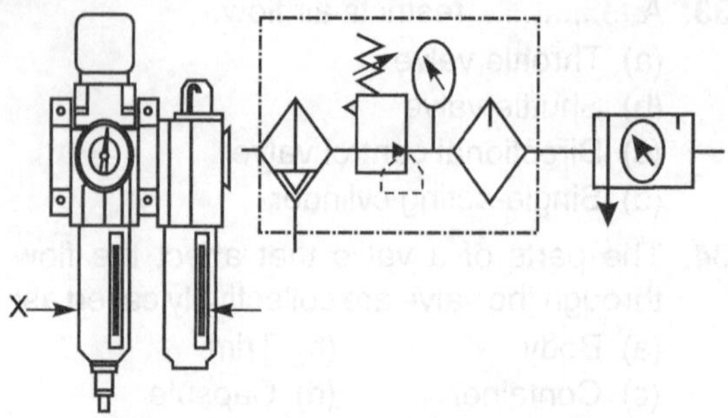

(a) Regulator (b) Control valve
(c) Filter (d) Air service unit

42. What is the name of the part marked 'X' in the figure given below?

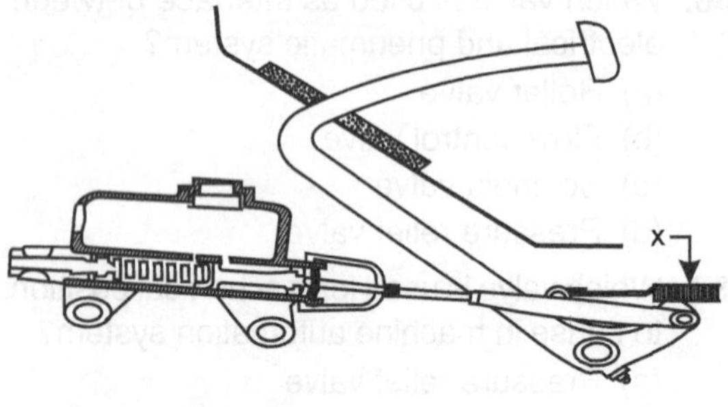

(a) Piston returning spring
(b) Brake shoe returning spring
(c) Brake pedal returning spring
(d) Pressure plate returning spring

43. What is the function of hydraulic oil in hydraulic system?
(a) To transmit power
(b) To increase heat
(c) To increase volume
(d) To reduce pressure

44. What symbol of control valve is shown in the figure given below?

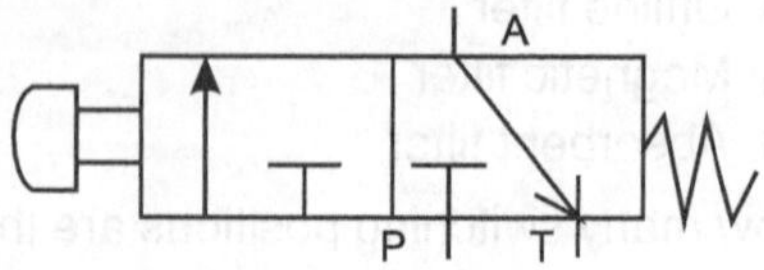

(a) 2/2 direction control valve
(b) 3/2 direction control valve
(c) 4/3 direction control valve
(d) 4/2 direction control valve

45. What is the sequence of power flow in a pneumatic system?
(a) Compressor – Actuator – Control valve
(b) Control valve – Compressor – Actuator
(c) Actuator – Control valve - Compressor
(d) Compressor – Control valve – Actuator

46. Which device transmits pressure equally to all wheel cylinders in hydraulic brake?
(a) Slave cylinder
(b) Servo cylinder
(c) Master cylinder
(d) Booster cylinder

47. Which component of air brake system converts pneumatic into mechanical energy?
(a) Reservoir (b) Brake valve
(c) Compressor (d) Brake chamber

48. What symbol of control valve is shown in the figure given below?

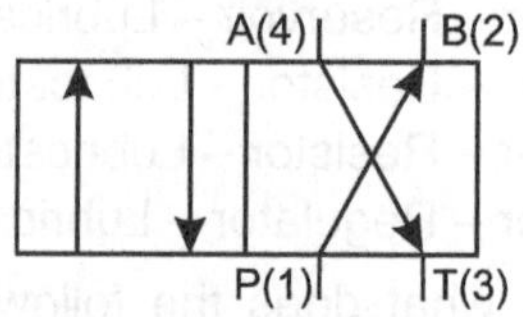

(a) 2/2 direction control valve
(b) 3/2 direction control valve
(c) 4/3 direction control valve
(d) 4/2 direction control valve

49. What symbol of control valve is shown in the figure given below?

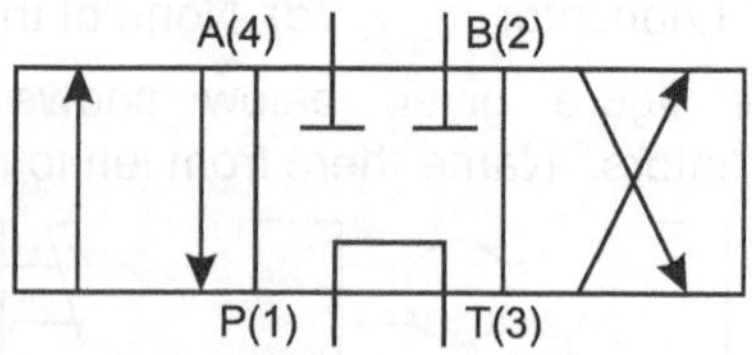

(a) 2/2 direction control valve
(b) 3/2 direction control valve
(c) 4/3 direction control valve
(d) 4/2 direction control valve

50. What is the type of inner drive gear used in the internal gear oil pump?
(a) Spur gear (b) Spiral gear
(c) Rotor gear (d) Helical gear

51. What causes failure of pneumatic brake?
(a) Less pedal play

(b) More pedal play
(c) Defective unloader valve
(d) Defective front wheel slack adjuster

52. What is the effect if the unloader valve is in open position?

(a) Air is released to the brake valve
(b) Air flows into the air tank
(c) Required air pressure is maintained
(d) Air is allowed continuously to the atmosphere

ANSWERS

1. (b)	**2.** (b)	**3.** (d)	**4.** (a)	**5.** (c)	**6.** (c)	**7.** (c)	**8.** (d)	**9.** (d)	**10.** (c)
11. (d)	**12.** (b)	**13.** (c)	**14.** (b)	**15.** (c)	**16.** (a)	**17.** (c)	**18.** (b)	**19.** (d)	**20.** (d)
21. (a)	**22.** (d)	**23.** (c)	**24.** (b)	**25.** (a)	**26.** (d)	**27.** (b)	**28.** (a)	**29.** (d)	**30.** (a)
31. (a)	**32.** (b)	**33.** (a)	**34.** (b)	**35.** (c)	**36.** (c)	**37.** (b)	**38.** (a)	**39.** (c)	**40.** (b)
41. (c)	**42.** (c)	**43.** (a)	**44.** (b)	**45.** (d)	**46.** (c)	**47.** (d)	**48.** (c)	**49.** (d)	**50.** (a)
51. (c)	**52.** (d)								

EXPLANATORY NOTES

3: The compressor is the most common industrial energy supply unit that converts mechanical energy into pneumatic energy. Pneumatic actuators are output devices for conversion of energy contained in compressed air to produce linear or rotary motion or apply a force.

4: A linear actuator, as used in hydraulic system, converts hydraulic power into a controllable linear force and/ormotion.

5: A double rod cylinder is shown in the figure given below.

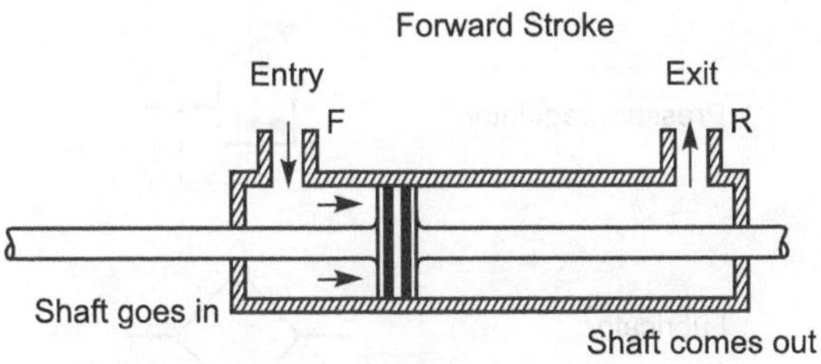

Double rod cylinder

6: A check valve permits flow through it in one direction and blocks the flow in the reverse direction.

7: A throttle valve is a device with a restriction that offers resistance to the system oil flowing through it. The throttle valve regulates the flow rate of the system oil. According to the type of restriction, throttle valves are of two types. They are: (1) Fixed type and (2) Adjustable type. In a fixed type throttle valve, the restriction is fixed, whereas in an adjustable type throttle valve, the area of the restriction can be varied.

9: Viscosity is a physical property of fluids. It shows resistance to flow.

10: Pascal's Law states that a pressure change at any point in a confined incompressible fluid is transmitted throughout the fluid such that the same change occurs everywhere.

13: A few examples of valve actuators' symbols are shown below.

The Valve Actuator Moves the Spool to its Various Positions:

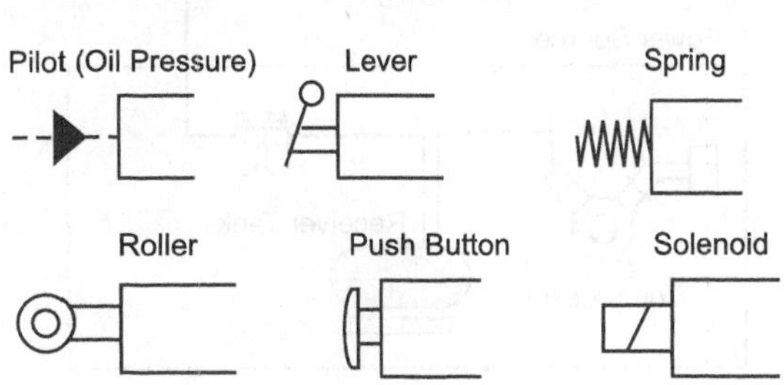

Valve actuators

24: The clutch slave cylinder plays an

important role in vehicles with manual transmissions. The slave cylinder works in tandem with the clutch master cylinder to disengage the clutch and shift the transmission when the pedal is pressed.

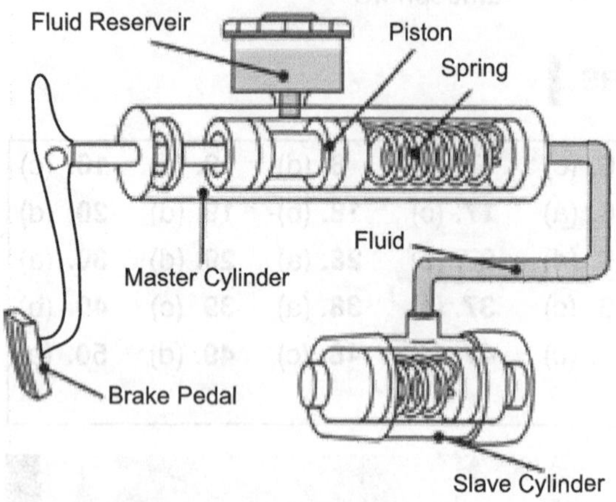

Slave cylinder in hydraulic brake system

25: A basic pneumatic system consists of three main blocks: (1) Power source, (2) Control valves and (3) Actuators. A typical pneumatic system with a number of components is depicted in the figure given below. The power source includes compressor, receiver tank,FRL etc.

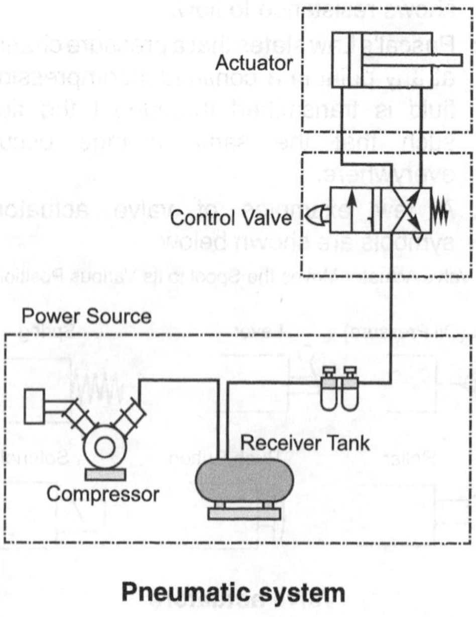

Pneumatic system

26: See crescent – shaped seal in internal gear pump in the figure given below.

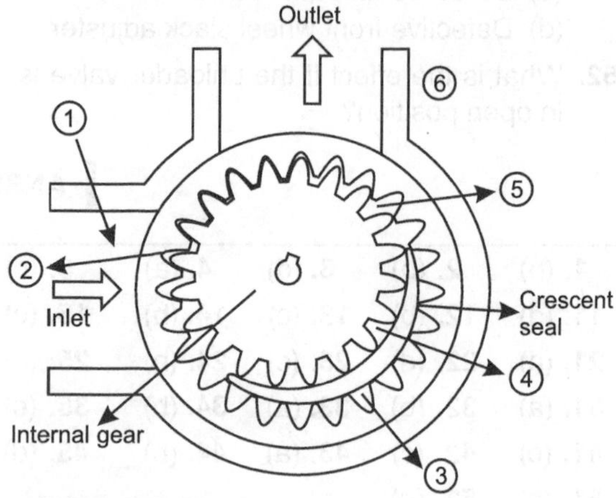

1. Oil enters here
2. Due to the vacuum caused by the un meshing of the teeth of these gears
3. Oil is carried in these spaces
4. And these spaces....
5. To this point, where constant meshing of two gears forces oil...
6. Out through this port...

Internal gear pump

30: Three units of FRL are shown in figure given below.

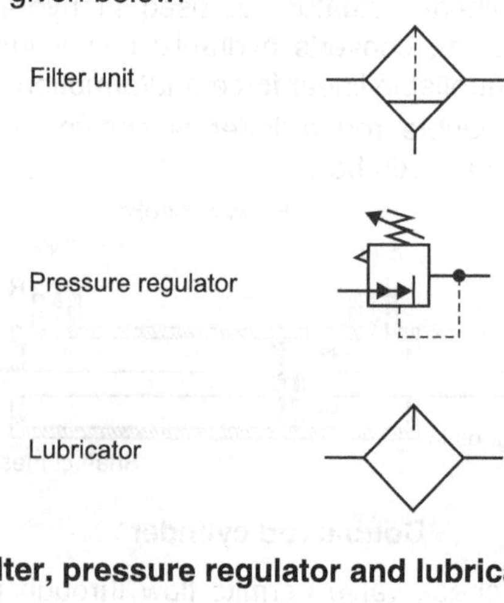

Filter, pressure regulator and lubricator

31: Symbols of various actuators are shown in figure given below.

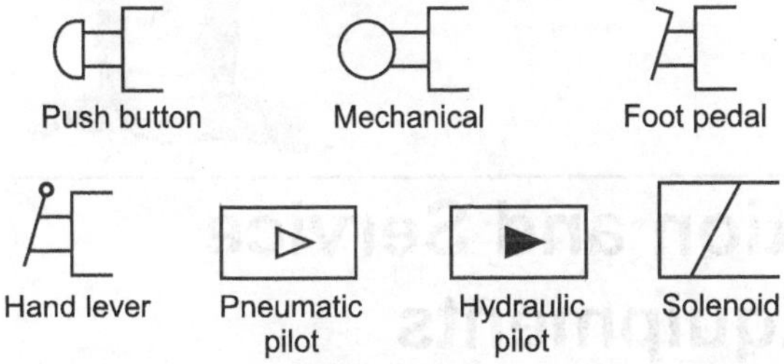

Actuators

34: The trim consists of the parts of the valve that affect the flow through the valve. In a standard globe valve the trim would just be the plug and seat. In a special valve the trim would consist of the plug, seat and retainer (or disk stack).

36: A solenoid valve is an electro mechanically operated valve. Solenoid valves differ in the characteristics of the electric current they use, the strength of the magnetic field they generate, the mechanism they use to regulate the fluid and the type and characteristics of fluid they control.

37: Pneumatic roller lever valves are used for mechanical position sensing in machine automation systems.

41: Compressed air in a pneumatic system is liable to be contaminated to a high degree. It is essential to remove fine dirt particles, to regulate the pressure, and perhaps to introduce a fine mist of oil in the compressed air to aid lubrication. These important functions can be accomplished through auxiliaryairline equipment, namely, filter, regulator and lubricator (FRL).

47: The components of air brake system are shown in the figure given below.

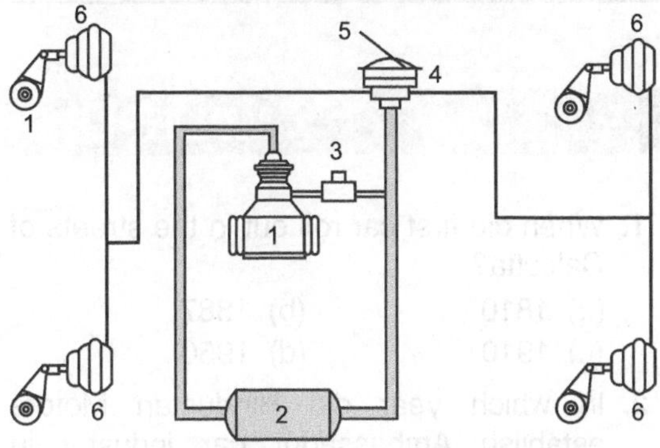

1. Compressor
2. Air reservoir
3. Governor
4. Brake valve
5. Foot pedal
6. Brake chamber
7. Slack adjuster

Air brake system

51: Unloader valve is mounted between the compressor and the reservoir to control the pressure of air in the reservoir.

Module 7

Specification and Service Equipments

MULTIPLE CHOICE QUESTIONS

1. When did first car roll out in the streets of Calcutta?

(a) 1810 (b) 1887
(c) 1910 (d) 1950

2. In which year did Hindustan Motors establish Ambassador car industry in India?

(a) 1900 (b) 1920
(c) 1940 (d) 1980

3. When was Maruti car launched in India?

(a) 1920 (b) 1940
(c) 1983 (d) 1990

4. What does 2498 CC with reference to vehicle specification indicate?

(a) Engine capacity
(b) Full tank capacity
(c) Single bore capacity
(d) Master cylinder capacity

5. What does the term '2WD' with reference to vehicle specification mean?

(a) Two-wheel drive
(b) Four-wheel drive
(c) Rear-wheel drive
(d) Front-wheel drive

6. Which of these increases the torque in the steering system?

(a) Drop arm (b) Gearbox
(c) Fluid pump (d) Knuckle arm

7. What information is given by '6J × 15'?

(a) Rim size (b) Tyre size
(c) Engine size (d) Vehicle size

8. Which digit indicates the engine type in VDC17 digit of VIN number?

(a) 2 (b) 3
(c) 5 (d) 8

9. What is indicated by '12 – 17 digit' of 17-digit VIN number?

(a) Body type (b) Vehicle type
(c) Serial number (d) Plant of product

10. What is the name of the component shown in the figure given below?

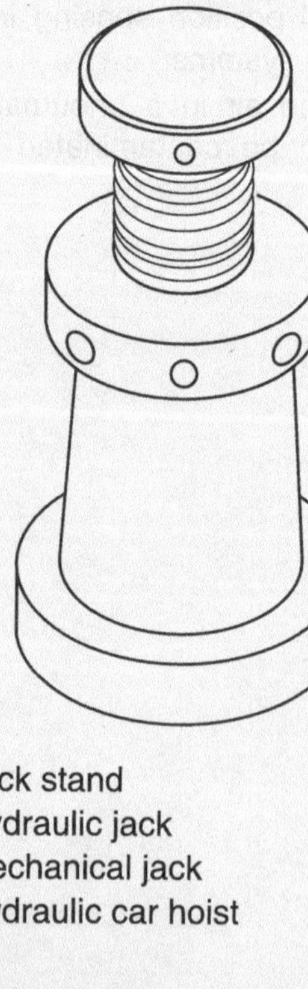

(a) Jack stand
(b) Hydraulic jack
(c) Mechanical jack
(d) Hydraulic car hoist

11. What is the name of the equipment shown in the figure given below?

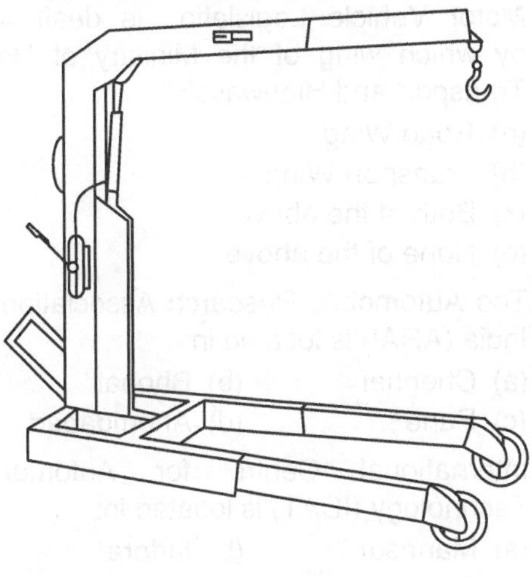

(a) Car hoist (b) Engine hoist
(c) Hydraulic jack (d) Hydraulic press

12. What is the name of the equipment shown in the figure given below?

(a) Car hoist (b) Grease gun
(c) Mechanical jack (d) Hydraulic jack

13. Which type of vehicle is capable of sensing its environment and navigating without human input?
(a) Diesel car (b) Petrol car
(c) Electric car (d) Driverless car

14. Which type of service equipment is used to hold the lifted vehicle for a long period?
(a) Hoist (b) Jack stand
(c) Arbor press (d) Hydraulic jack

15. Which type service equipment is used in car service station to lift the car?
(a) Arbor press (b) Hydraulic jack
(c) Hydraulic hoist (d) Hydraulic press

16. What is used for quick inspection under the chassis of a car?
(a) Stands (b) Screw jack
(c) Trolley jack (d) Two-post hoist

17. Tachometer in a diesel engine measures:
(a) Speed
(b) Distance covered
(c) Engine rpm
(d) Fuel consumption

18. What is used to check oil level in the engine sump?
(a) Feeler gauge (b) Plastic gauge
(c) Oil pressure gauge (d) Dip stick

19. Which mobile hoist is used for self loading and shifting from one place to another within a factory?
(a) Jib crane (b) Fork lift
(c) Chain hoist (d) Gear hoist

20. In small garages, which type of jet nozzle is used in water washing machine?
(a) Mechanized jet type
(b) Manual jet type
(c) Air jet type
(d) Hydraulic jet type

21. A grease gun is used to lubricate the grease points periodically. In a large service station, which type of grease gun is used?
(a) Mechanical grease gun
(b) Hydraulic grease gun
(c) Electric grease gun
(d) Pneumatic grease gun

22. Which one of the following is not in the trade of buses?
(a) Ashok Leyland (b) Tata
(c) Force (d) HMT

23. Oil level in an engine should be checked:
(a) Before starting the engine
(b) During running of the engine
(c) Within two minutes after stopping the engine
(d) Any time after starting the engine

24. A vehicle is classified as Heavy Commercial Vehicle, if the Gross Vehicle Weight (GVW) is:
(a) Not more than 4.2 metric tonne
(b) More than 4.2 metric tonne but less than 7.5 metric tonne
(c) More than 7.5 metric tonne, but less than 16.2 metric tonne
(d) More than 16.2 metric tonne

25. Mini cars have length up to:
(a) 3400 mm (b) 4000 mm
(c) 4500 mm (d) 4700 mm

26. Utility vehicles have a seating capacity of at least .... persons besides the driver.
(a) Five (b) Seven
(c) Nine (d) Eleven

27. According to SIAM classification, Alto, Wagon R, Zen, i10, A-Star, Swift, Palio, Indica cars fall in the segment:
(a) Mini (b) Compact
(c) Midsize (d) Executive

28. Maruti 800, Alto, Santro and i10 cars are examples of:
(a) Van (b) Hatchback
(c) Sedan (d) Estate

29. Which of these is NOT a two-wheeler?
(a) Motorcycle (b) Scooter
(c) Moped (d) Mini Van

30. Motor Vehicle Legislation is dealt with by which wing of the Ministry of Road Transport and Highways?
(a) Road Wing
(b) Transport Wing
(c) Both of the above
(d) None of the above

31. The Automobile Research Association of India (ARAI) is located in:
(a) Chennai (b) Bhopal
(c) Pune (d) Ahemdabad

32. International Centre for Automotive Technology (ICAT) is located in:
(a) Manesar (b) Indore
(c) Chennai (d) Silchar

33. When raising or lowering a hoist, who is responsible to ensure no one is under the vehicle?
(a) Shop foreman
(b) Instructor
(c) Safety coordinator
(d) Person pushing the button

34. Which of the following are used to support a vehicle after it has been raised?
(a) Floor jacks
(b) Jack stands
(c) Transmission jacks
(d) Bumper lifts

35. Which of the following workplace accidents may be the direct result of the improper use of a floor jack?
(a) Mechanical damage to vehicle
(b) Personal injury
(c) Death
(d) All of the above

ANSWERS

1. (b)	**2.** (c)	**3.** (c)	**4.** (a)	**5.** (a)	**6.** (b)	**7.** (a)	**8.** (d)	**9.** (c)	**10.** (c)
11. (b)	**12.** (d)	**13.** (d)	**14.** (a)	**15.** (c)	**16.** (c)	**17.** (c)	**18.** (d)	**19.** (b)	**20.** (b)
21. (d)	**22.** (d)	**23.** (a)	**24.** (d)	**25.** (a)	**26.** (b)	**27.** (b)	**28.** (b)	**29.** (d)	**30.** (b)
31. (c)	**32.** (a)	**33.** (d)	**34.** (b)	**35.** (d)					

EXPLANATORY NOTES

6: The steering gear box converts the rotary movement of the steering wheel into a straight line movement of the steering linkage; it also makes it easier for the driver to steer by giving a gear reduction.

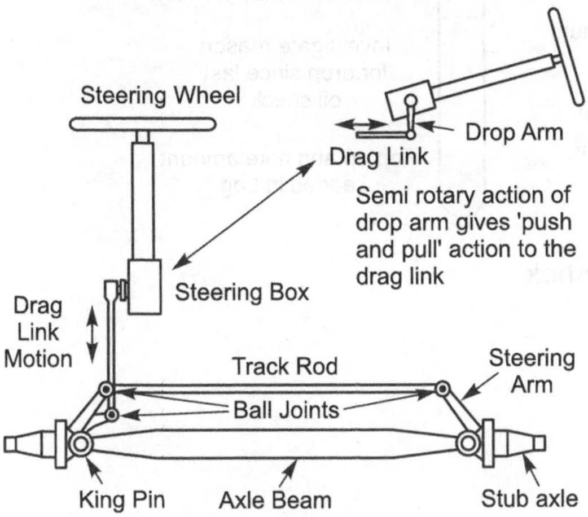

Steering system

8: The car's vehicle identification number (VIN) is the identifying code for a SPECIFIC automobile. The VIN serves as the car's fingerprint, as no two vehicles in operation have the same VIN. A VIN is composed of 17 characters (digits and capital letters) that act as a unique identifier for the vehicle.

Vehicle Identification Number (VIN) is the legal identity of your vehicle. The vehicle identification number is stamped on the VIN plate riveted on to the floor. Lift the floor carpet tab in front of front passenger seat to access VIN plate.

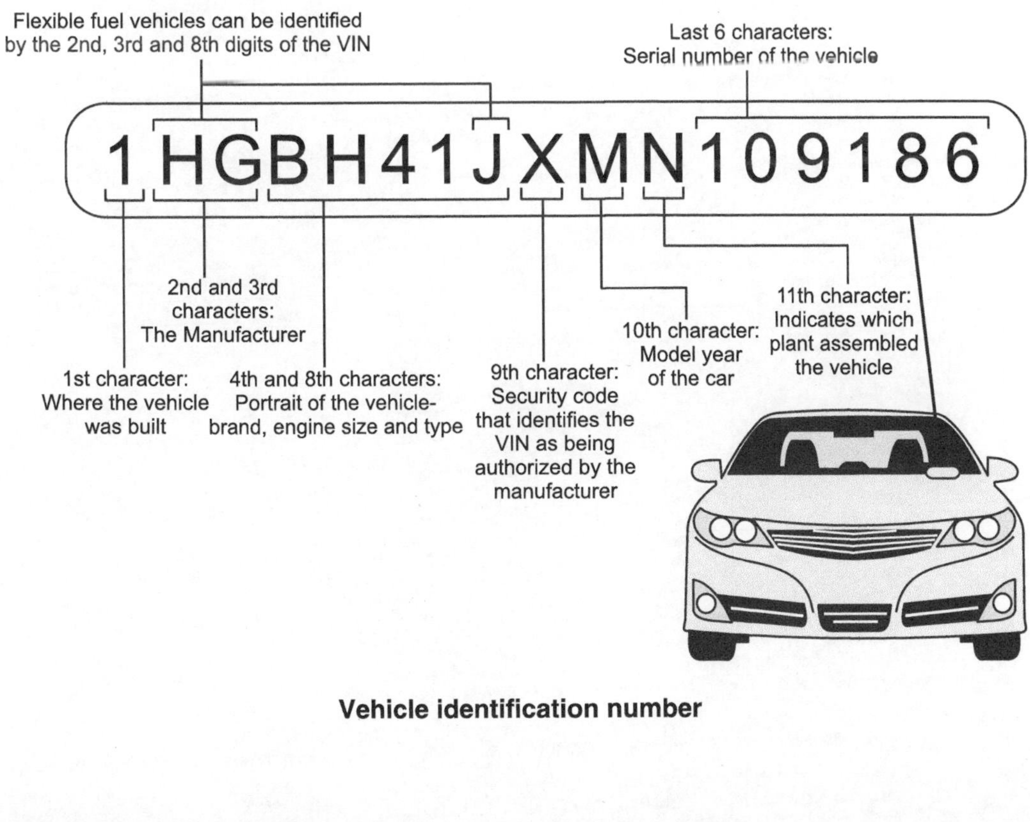

Vehicle identification number

9: See figure under explanatory note of MCQ 8.

17: A tachometer is a sensor device used to measure the rotation speed of an object such as the engine shaft in a car, and is usually restricted to mechanical or electrical instruments. This device indicates the revolutions per minute (RPM) performed by the object.

18: The use of dip stick to check oil level is illustrated in the figure given below.

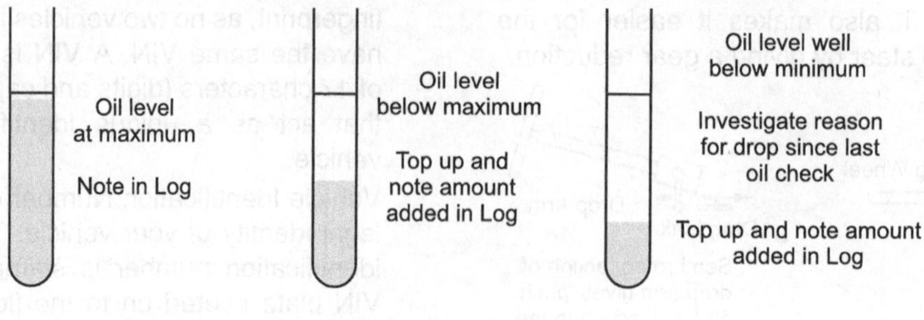

Use of dip stick

Module 8

Diesel Engine Overview

MULTIPLE CHOICE QUESTIONS

1. What is the type of engine shown in the figure given below?

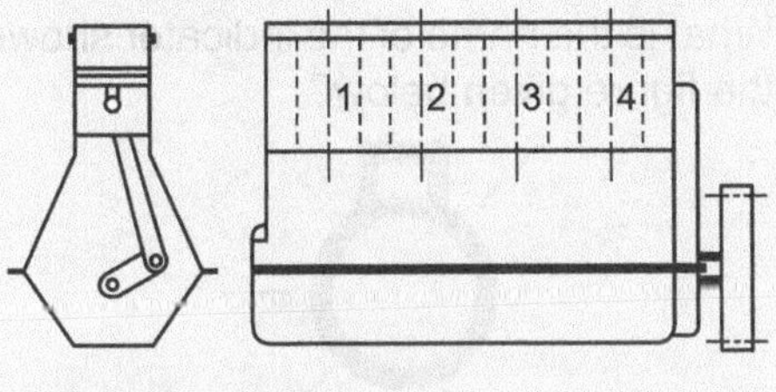

(a) Inline engine (b) V engine
(c) Opposed engine (d) Radial engine

2. What is the type of engine shown in the figure given below?

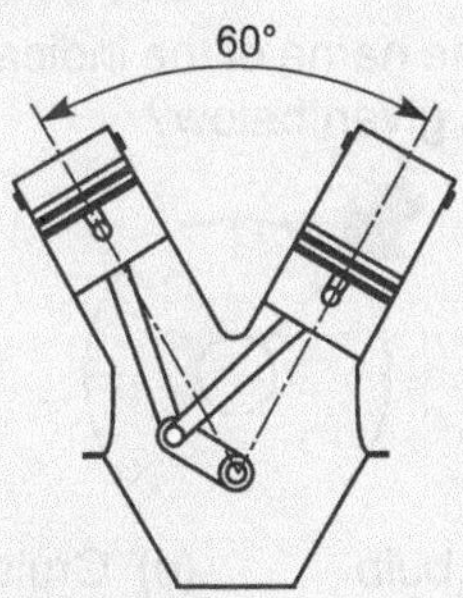

(a) Inline engine (b) V engine
(c) Opposed engine (d) Radial engine

3. What is the type of engine shown in the figure given below?

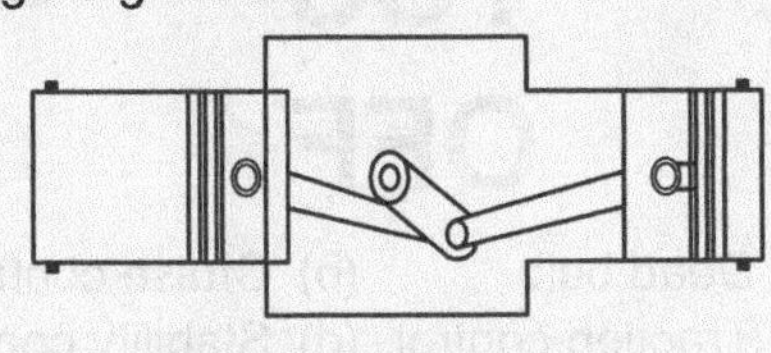

(a) Inline engine (b) V engine
(c) Opposed engine (d) Radial engine

4. What is the type of engine shown in the figure given below?

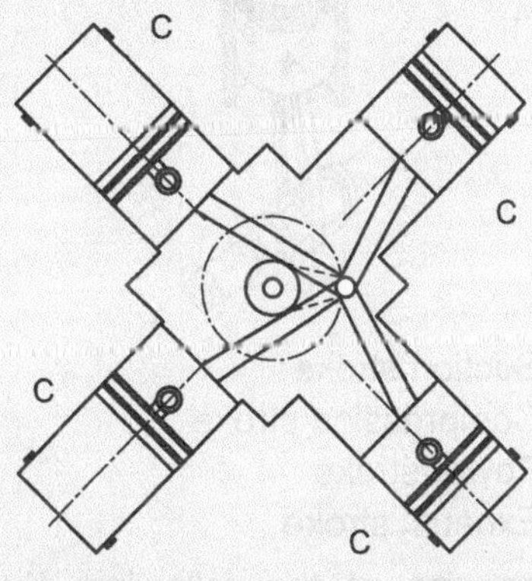

(a) Inline engine (b) V engine
(c) Opposed engine (d) Radial engine

5. Which one of the following types of engines has the longest crankshaft?

(a) Opposed engine
(b) Radial engine
(c) Inline engine
(d) V engine

6. Identify the stroke as shown in the figure given below.

(a) Compression stroke
(b) Power stroke
(c) Suction stroke
(d) Exhaust stroke

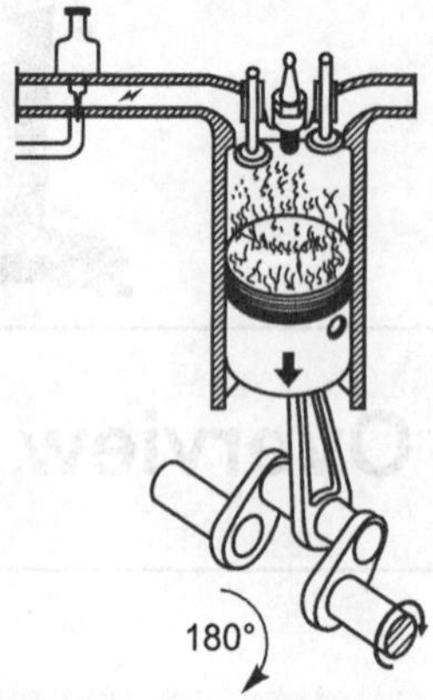

7. Identify the stroke as shown in the figure given below.

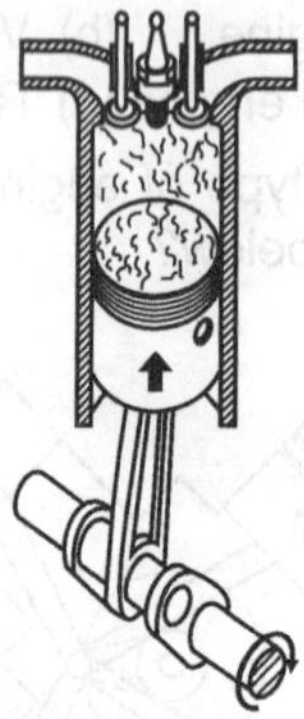

(a) Suction stroke
(b) Compression stroke
(c) Power stroke
(d) Exhaust stroke

8. Which one of the following gives the correct position of inlet and exhaust valves during power stroke?
(a) Inlet valve is open and exhaust valve is closed
(b) Exhaust valve is open and inlet valve is closed
(c) Both inlet and exhaust valves are in closed position
(d) Both inlet and exhaust valves are in open position

9. A four-stroke engine produces one power stroke in:
(a) 2 revolutions of crankshaft
(b) 4 revolutions of crankshaft
(c) 6 revolutions of crankshaft
(d) 8 revolutions of crankshaft

10. A two-stroke engine produces one power stroke in:
(a) 1revolution of crankshaft
(b) 2 revolutions of crankshaft
(c) 3 revolutions of crankshaft
(d) 4 revolutions of crankshaft

11. In an engine, stroke length is equal to:
(a) Half the throw of crank
(b) Equal to throw of crank
(c) Double the throw of crank
(d) Three times the throw of crank

12. The stroke of an engine is:
(a) Inside diameter of the cylinder
(b) Length of the cylinder
(c) Distance between t.d.c. and b.d.c.
(d) Length of the crankshaft

13. What is the name of the indicator shown in the figure given below?

(a) Dead bulb (b) Cruise control
(c) Traction control (d) Stability control

14. What is the name of the indicator shown in the figure given below?

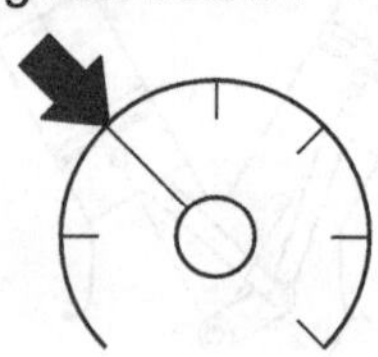

(a) Dead bulb (b) Cruise control
(c) Traction control (d) Stability control

15. What is the name of the indicator shown in the figure given below?

TCS
OFF

(a) Dead bulb (b) Cruise control
(c) Traction control (d) Stability control

16. What is the name of the indicator shown in the figure given below?

(a) Centre differential lock
(b) Proximity sensor
(c) Economy mode
(d) Electric power steering

17. What is the name of the indicator shown in the figure given below?

(a) Seat belt indicator
(b) Air bag indicator
(c) Brake indicator
(d) ABS indicator

18. What is the name of the indicator shown in the figure given below?

(a) Seat belt (b) Air bag
(c) Brake (d) ABS

19. A four-cylinder engine has a capacity of 2.4 litre. The swept volume of one cylinder engine is:
(a) 400 cm^3 (b) 600 cm^3
(c) 1200 cm^3 (d) 2400 cm^3

20. An engine cylinder has clearance volume of 100 cm^3 and its swept volume is 800 cm^3. The compression ratio is:
(a) 10 : 1 (b) 9 : 1
(c) 8 : 1 (d) 7 : 1

21. Which one of the following gives the correct sequence of Otto cycle engine?
(a) Induction, power, compression, exhaust
(b) Induction, exhaust, compression, power
(c) Induction, compression, power, exhaust
(d) Induction, power, exhaust, compression

22. A 4-stroke engine makes 1000 revolutions per minute, the number of power stokes per minute will be:
(a) 250 (b) 500
(c) 750 (d) 1000

23. Which one of the following steps will result in increasing the stroke length?
(a) Piston made shorter
(b) Connecting rod lengthened
(c) Crankshaft throw increased
(d) Gudgeon pin moved nearer to the crankshaft

24. Which one of the following engine components '*carries*' the engine over its non-working strokes?
(a) Piston (b) Crankshaft
(c) Connecting rod (d) Flywheel

25. During which one of the following strokes, the pressure in the engine cylinder is less than the atmospheric pressure?
(a) Induction (b) Compression
(c) Power (d) Exhaust

26. The ratio between BHP and IHP is called:
(a) Thermal efficiency
(b) Engine efficiency
(c) Mechanical efficiency
(d) Power efficiency

27. A two-stroke engine has ports in the:
(a) Piston (b) Cylinder walls
(c) Piston rings (d) Crankcase

28. An engine should not be started in a closed garage, because:
(a) exhaust gas is poisonous
(b) of the risk of explosion
(c) the engine receives an incorrect mixture
(d) the engine is damaged if the gas cannot escape

29. If there is resistance to the flow of air during intake stroke, it will:
(a) Reduce power of the engine
(b) Increase power of the engine
(c) Have no effect on the power of the engine
(d) Cause fluctuation on power of the engine

30. What is drawn in the cylinder of diesel engine during suction stroke?
(a) A mixture of air and fuel
(b) Pure air alone
(c) Fuel alone
(d) Gas

31. In a diesel engine, fuel and air mix together in the:
(a) Carburettor
(b) Injector
(c) Combustion chamber
(d) Inlet port

32. In diesel engine, the fuel is ignited by:
(a) A glow plug
(b) A spark plug
(c) An injector
(d) Virtue of temperature of compressed air

33. What is compressed in the cylinder of a diesel engine during the compression stroke?
(a) Diesel
(b) Air
(c) Fuel
(d) Mixture of air and diesel

34. What engine starting system is shown in the figure given below?

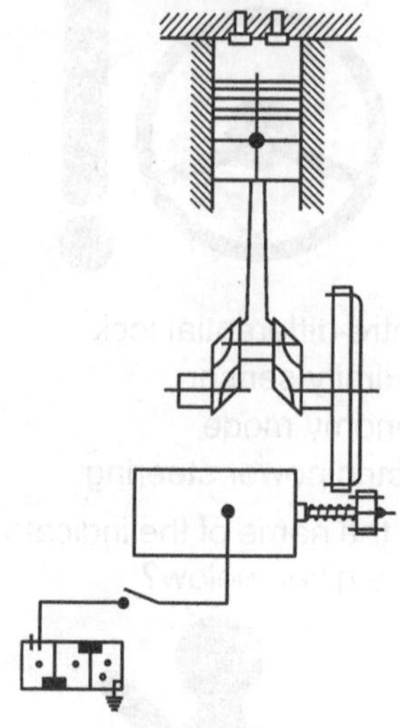

(a) Compressed air cranking
(b) Hydraulic cranking
(c) Electric motor cranking
(d) Gasoline engine cranking

35. What engine starting system is shown in the figure given below?

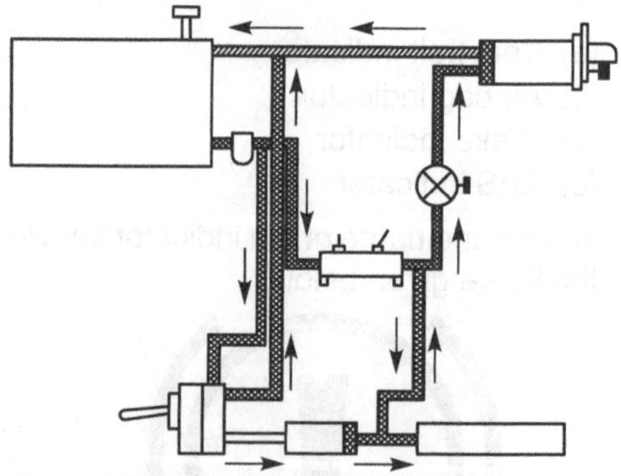

(a) Compressed air cranking
(b) Hydraulic cranking
(c) Electric motor cranking
(d) Gasoline engine cranking

36. What is the name of the part marked 'X' in the figure given below?

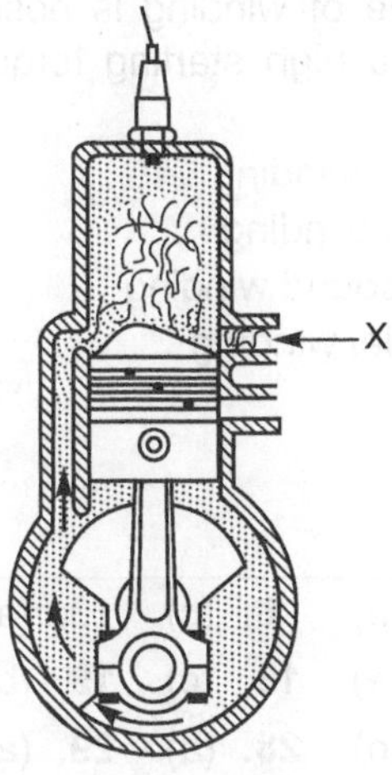

(a) Intake port (b) Exhaust port
(c) Transfer port (d) Injection port

37. Which engine has its cylinders at 60° to each other?
(a) Inline engine (b) 'V' engine
(c) Opposed engine (d) Radial engine

38. Which engine has four cylinders each at 90° to the next one?
(a) Inline engine
(b) 'V' engine
(c) Opposed engine
(d) Radial engine

39. Which type of engine is most preferred in aeroplanes?
(a) Inline engine (b) 'V' engine
(c) Opposed engine (d) Radial engine

40. Which engine is also called constant pressure cycle engine?
(a) Otto cycle (b) Diesel cycle
(c) Dual cycle (d) Rankine cycle

41. The distance between the centre of the crank pin to the centre of the main journal is called:
(a) Stroke (b) Cycle
(c) Throw (d) Bore

42. The piston stroke is defined as:
(a) Double the throw
(b) Half of the throw
(c) Equal to the throw
(d) 4 times the throw.

43. What is firing order of an engine?
(a) Sequence of power stroke occurrence
(b) Sequence of suction stroke occurrence
(c) Sequence of compression stroke occurrence
(d) Sequence of exhaust stroke occurrence

44. What is the use of bourdon gauge?
(a) To measure pressure
(b) To measure temperature
(c) To measure volume
(d) To measure heat

45. Which method is used to start a heavy earth moving machinery engine?
(a) Hand cranking
(b) Electric motor cranking
(c) Compressed air cranking
(d) Gasoline engine cranking

46. Which type of engines adopt hand cranking?
(a) Small engines
(b) LMV engines
(c) Heavy vehicle engines
(d) Stationary engines

47. Which is the most preferred starting system in LMV?
(a) Hand cranking
(b) Compressed air cranking
(c) Gasoline engine cranking
(d) Electric motor cranking

48. In an engine, piston stroke length and bore diameter are equal. This type of engine is called:
(a) Low speed engine
(b) Over-square engine
(c) Under-square engine
(d) Square engine

49. In an opposed piston engine, what is the location of the combustion chamber?
(a) Above the piston
(b) Below the piston
(c) In the space between the two pistons
(d) None of the above

50. What parameters decide the firing order in an I.C. engine?
(a) Arrangement of cylinders
(b) Design of crankshaft
(c) Number of cylinders
(d) All of the above

51. What type of winding is best suitable to produce a high starting torque in starter motor?
(a) Series winding
(b) Shunt winding
(c) Compound winding
(d) Parallel winding

ANSWERS

1. (a)	**2.** (b)	**3.** (c)	**4.** (d)	**5.** (c)	**6.** (c)	**7.** (b)	**8.** (c)	**9.** (a)	**10.** (a)
11. (c)	**12.** (b)	**13.** (a)	**14.** (b)	**15.** (c)	**16.** (d)	**17.** (a)	**18.** (c)	**19.** (b)	**20.** (b)
21. (c)	**22.** (b)	**23.** (c)	**24.** (d)	**25.** (a)	**26.** (c)	**27.** (b)	**28.** (a)	**29.** (a)	**30.** (b)
31. (c)	**32.** (d)	**33.** (b)	**34.** (c)	**35.** (b)	**36.** (b)	**37.** (b)	**38.** (d)	**39.** (d)	**40.** (b)
41. (c)	**42.** (a)	**43.** (a)	**44.** (a)	**45.** (d)	**46.** (a)	**47.** (d)	**48.** (d)	**49.** (c)	**50.** (d)
51. (a)									

EXPLANATORY NOTES

1: In inline type of engine, the cylinders are arranged in one line. The length of the crankshaft is longer than that of the other types of engines.

Inline engine: Better balancing and more uniform torque is in this type.

V engine: In this type, the engine height is lower than it is in the in-line engine.

Opposed engine: This provides better mechanical balance. This type of engine can run smoothly even at a much higher speed. It also gives higher output.

Radial engine: This type of engine is shorter, lighter and more rigid. It is used mostly in aeroplanes.

18: An anti-lock braking system(ABS) is a safety anti-skid braking system. ABS operates by preventing the wheels from locking up during braking, thereby maintaining tractive contact with the road surface.

20: Compression ratio = (Swept volume + Clearance volume) / Clearance volume

34: In electric motor cranking system, a starter motor is used to rotate the flywheel of the engine. A battery is used to supply power to the starter motor.

35: In hydraulic cranking system, hydraulic fluid under pressure passes through hydraulic starter motor to rotate the engine flywheel.

40: An **Otto cycle** is an idealized thermodynamic cycle that describes the functioning of a typical spark ignition piston engine. In Otto cycle engine, combustion takes place at constant volume. The **Diesel cycle** is a combustion process of a reciprocating internal combustion engine. In it, fuel is ignited by heat generated during the compression of air in the combustion chamber, into which fuel is then injected. This is in contrast to igniting the fuel-air mixture with a spark plug as in the Otto cycle engine. The **dual combustion cycle** is a thermal cycle that is a combination of the Otto cycle and the Diesel cycle. The Rankine cycle is an idealized thermodynamic cycle of a heat engine that converts heat into mechanical work while undergoing phase change.

45: Gasoline engine starting is used to start heavy earth moving machinery engines. Starting of the gasoline engine is done either by hand cranking or by an electric motor. The gasoline engine then cranks the heavy machinery engine.

49: Opposed piston engine and location of combustion chamber is shown in the figure given below.

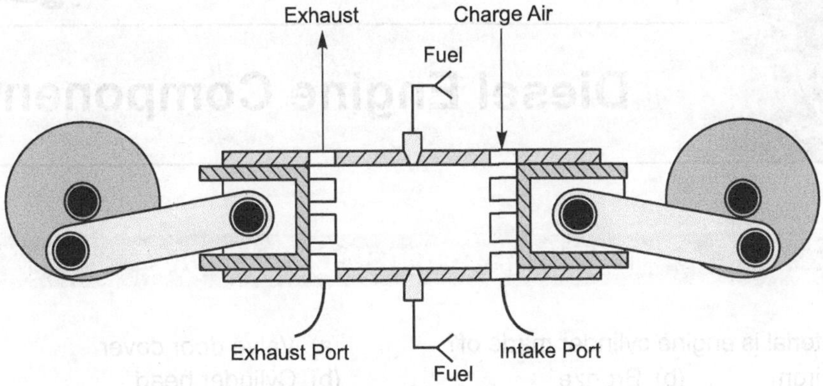

Combustion chamber in opposed piston engine

Module 9

Diesel Engine Components

MULTIPLE CHOICE QUESTIONS

1. What material is engine cylinder made of?

(a) Cast iron (b) Bronze

(c) Brass (d) Zinc alloy

2. What is the name of the part marked 'X' in the figure given below?

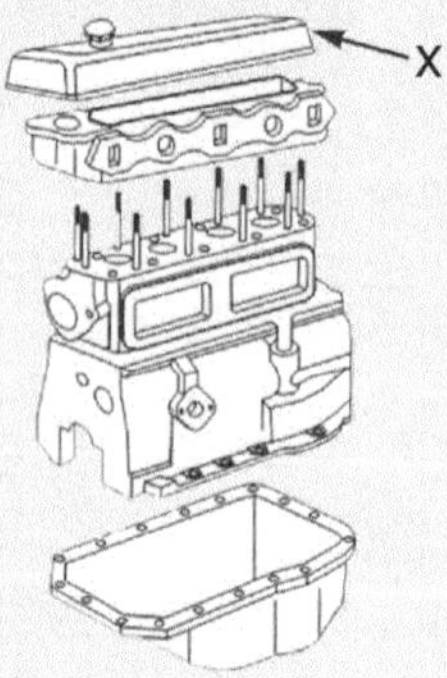

(a) Valve door cover

(b) Cylinder head

(c) Engine block

(d) Oil pan

3. What is the name of the part marked 'X' in the figure given below?

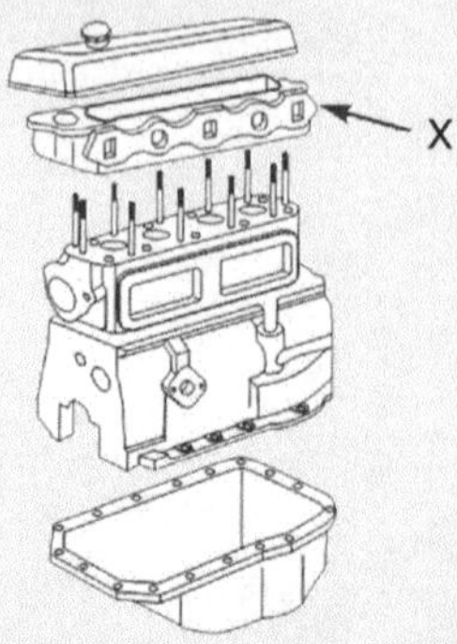

a) Valve door cover

(b) Cylinder head

(c) Engine block

(d) Oil pan

4. What is the name of the diagram shown below?

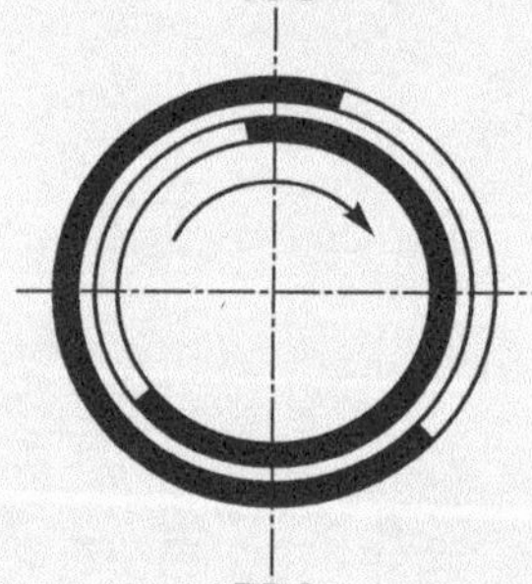

(a) Injection timing

(b) Port timing

(c) Valve timing

(d) Ignition timing

5. What is the name of the part marked 'X' in the figure given below?

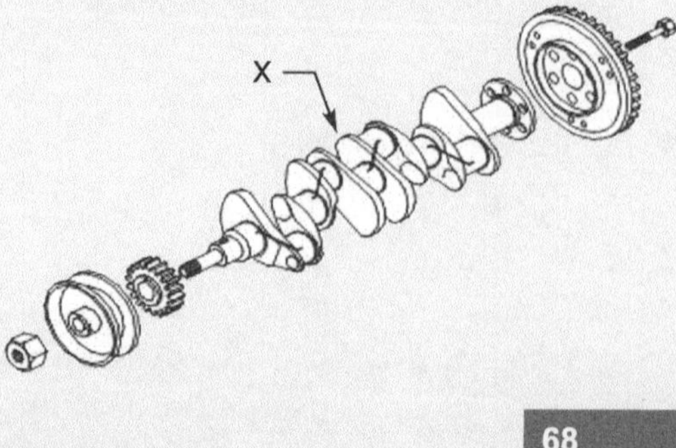

(a) Crank pin (b) Crank arm
(c) Balancing weight (d) Main journal

6. Identify what is marked 'X' in the figure given below.

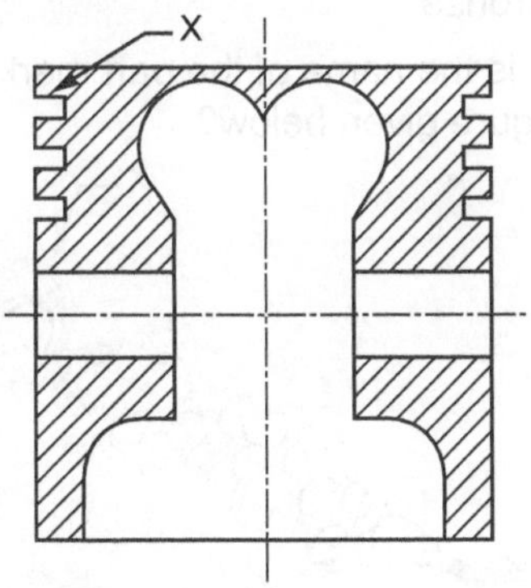

(a) Ring land
(b) Gudgeon pin boss
(c) Ring grooves
(d) Crown

7. Identify the section of a piston marked 'X' in the figure given below.

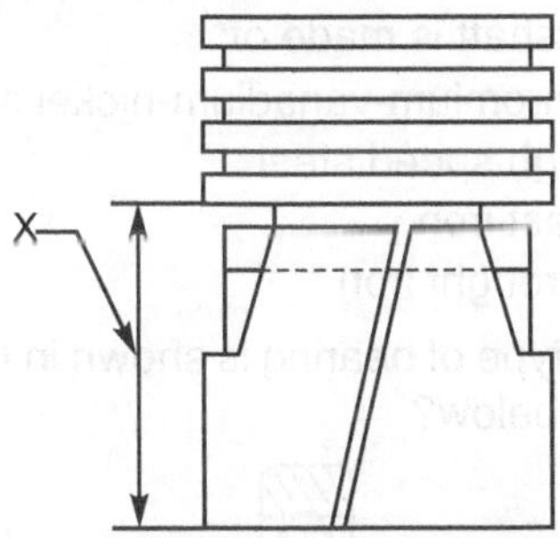

(a) Skirt section (b) Crown section
(c) Ring section (d) Face section

8. What is the name of the part marked 'X' in the figure given below?

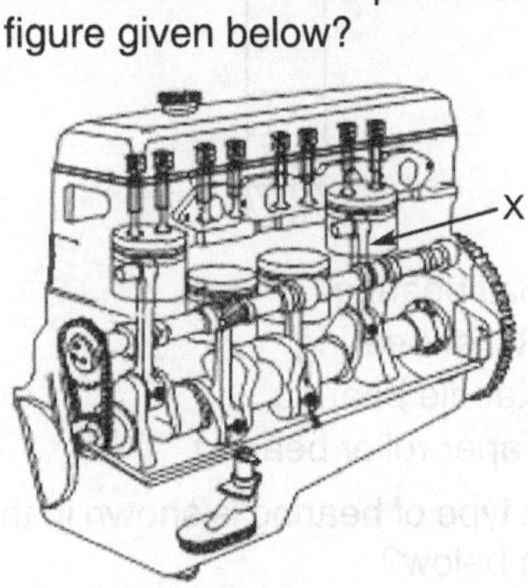

(a) Piston (b) Camshaft
(c) Crankshaft (d) Connecting rod

9. What is the name of the part marked 'X' in the figure given below.

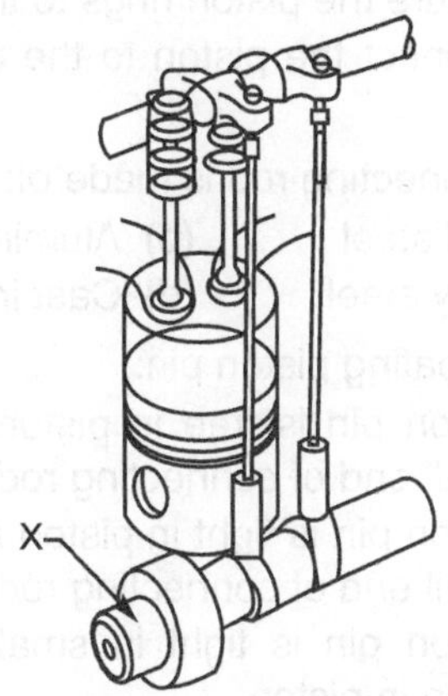

(a) Cylinder head (b) Engine block
(c) Connecting rod (d) Cam shaft

10. What is the type of piston head shown in the figure given below?

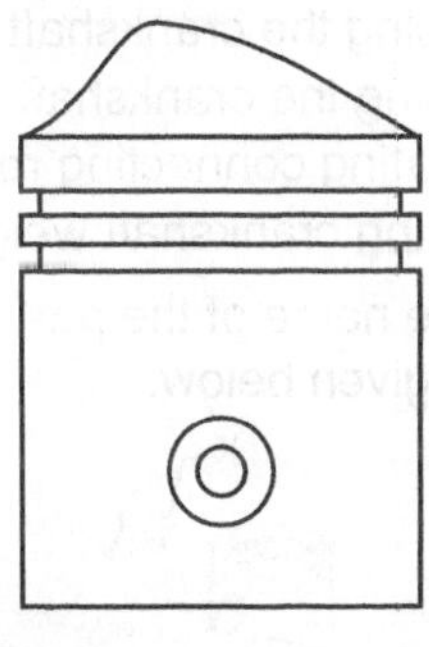

(a) Flat head (b) Domed head
(c) Concave head (d) Irregular head

11. Piston rings are generally made of:
(a) Brass (b) Copper
(c) Cast iron (d) Aluminium

12. Which part of the piston is subjected to high pressure and temperature?
(a) Crown (b) Skirt
(c) Land (d) Ring section

13. The purpose of piston rings is to control:
(a) Combustion pressure
(b) Cylinder wall lubrication
(c) Oil consumption
(d) All of the above

14. The purpose of the gudgeon pin is to:
(a) Prevent the valve from rotating
(b) Link the connecting rod to the crankshaft
(c) Secure the piston rings to the piston
(d) Connect the piston to the connecting rod

15. The connecting rod is made of:
(a) Mild steel (b) Aluminium
(c) Alloy steel (d) Cast iron

16. In full floating piston pin:
(a) Piston pin is free in piston bore and small end of connecting rod
(b) Piston pin is tight in piston and free in small end of connecting rod
(c) Piston pin is tight in small end and loose in piston
(d) Piston pin is tight in both (bore and small end)

17. A hole is drilled between the crankshaft main journal and crank pin for:
(a) Balancing the crankshaft
(b) Reducing the crankshaft vibrations
(c) Lubricating connecting rod bearings
(d) Reducing crankshaft weight

18. What is the name of the part marked 'X' in the figure given below.

(a) Gudgeon pin (b) King pin
(c) Connecting rod (d) Piston pin hole

19. What material piston pins are made of?
(a) Nickel – chromium steel
(b) Cast iron
(c) HSS
(d) Bronze

20. What is the name of the part marked 'X' in the figure given below?

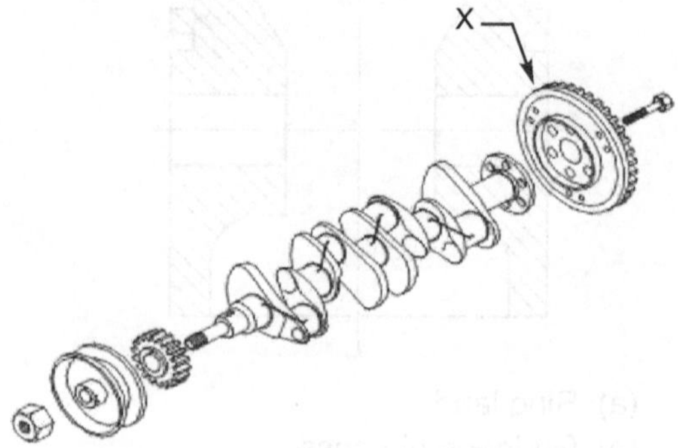

(a) Drive wheel
(b) Vibration damper
(c) Fan pulley
(d) Flywheel

21. Crankshaft is made of:
(a) Chromium-vanadium-nickel steel
(b) High speed steel
(c) Cast iron
(d) Wrought iron

22. What type of bearing is shown in the figure given below?

(a) Ball bearing
(b) Roller bearing
(c) Needle bearing
(d) Taper roller bearing

23. What type of bearing is shown in the figure given below?

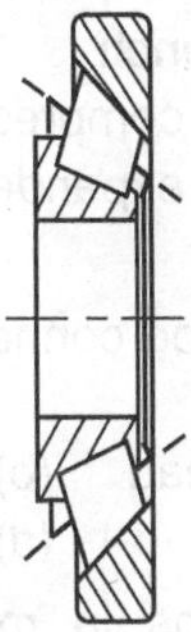

(a) Ball bearing
(b) Roller bearing
(c) Needle bearing
(d) Taper roller bearing

24. The part of piston below piston boss is called:
(a) Land of the piston
(b) Ring of the piston
(c) Crown of the piston
(d) Skirt of the piston

25. What connects the piston with the crankpin?
(a) Push rod (b) Connecting rod
(c) Camshaft (d) Crankshaft

26. What is the name of the part marked 'X' in the figure given below?

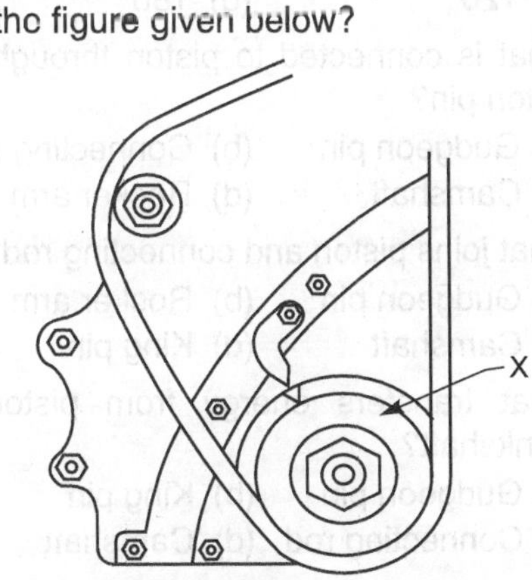

(a) Alternator pulley
(b) Water pump pulley
(c) Crankshaft pulley
(d) Dynamo pulley

27. What is the type of piston head shown in the figure given below?

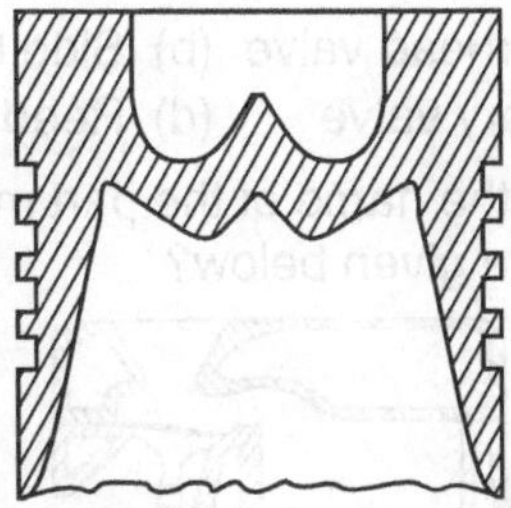

(a) Flat head (b) Domed head
(c) Concave head (d) Irregular head

28. What is the type of piston head shown in the figure given below?

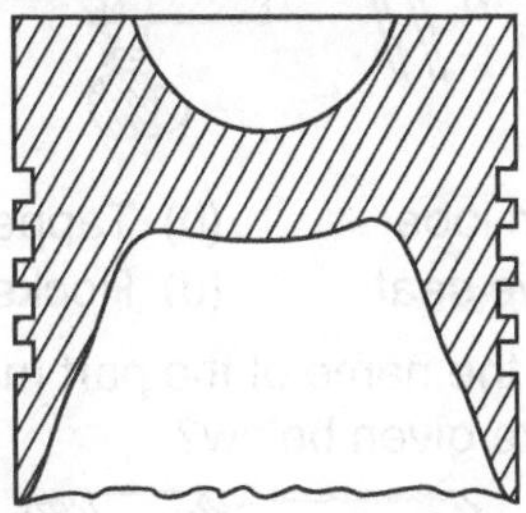

(a) Flat head (b) Domed head
(c) Concave head (d) Irregular head

29. What tool is used to remove the crankshaft pulley?
(a) Double – ended spanner
(b) Ring spanner
(c) Pipe wrench
(d) Puller

30. Inlet valve is made of:
(a) Nickel steel alloy
(b) Aluminium alloy
(c) Copper alloy
(d) Chromium alloy

31. What type of valve mechanism is shown in the figure given below?

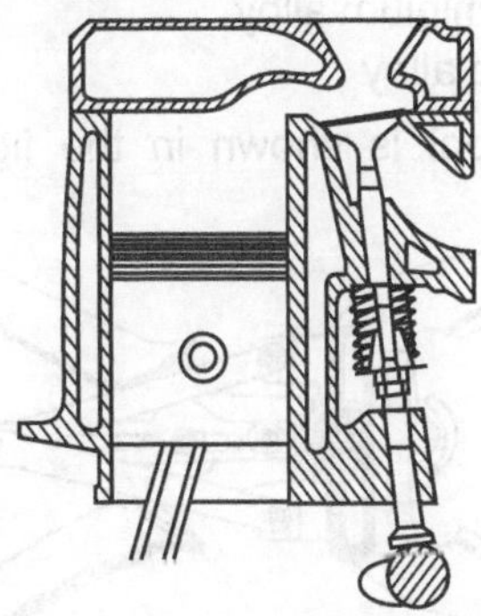

(a) Overhead valve (b) Side head valve
(c) Rotary valve (d) Reed valve

32. What is the name of the part marked 'X' in the figure given below?

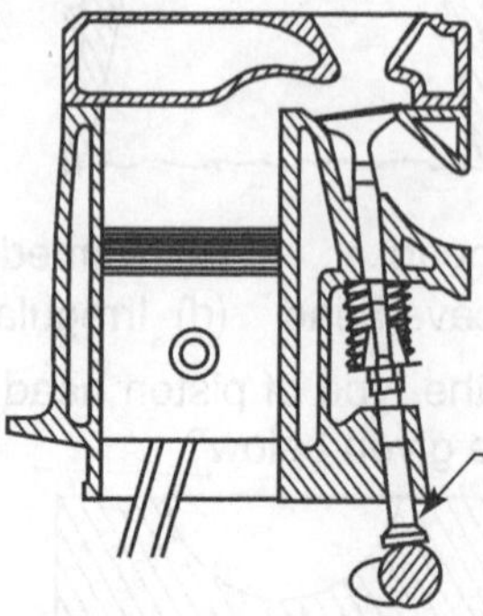

(a) Cam lobe (b) Tappet
(c) Valve seat (d) Rocker arm

33. What is the name of the part marked 'X' in the figure given below?

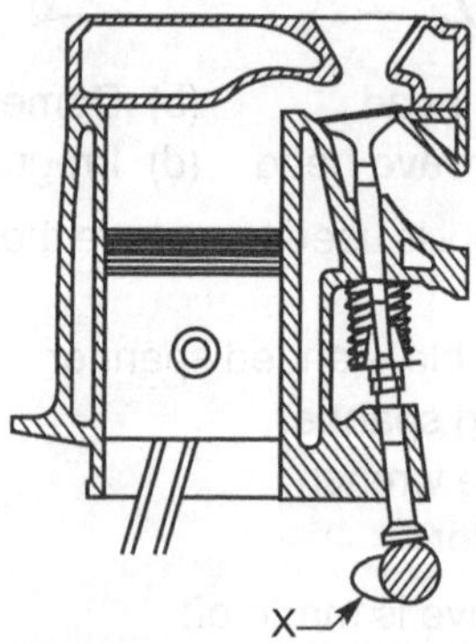

(a) Cam lobe (b) Tappet
(c) Valve seat (d) Rocker arm

34. Camshaft is made of:
(a) Forged alloy steel
(b) Copper alloy
(c) Aluminium alloy
(d) Zinc alloy

35. What tool is shown in the figure given below?

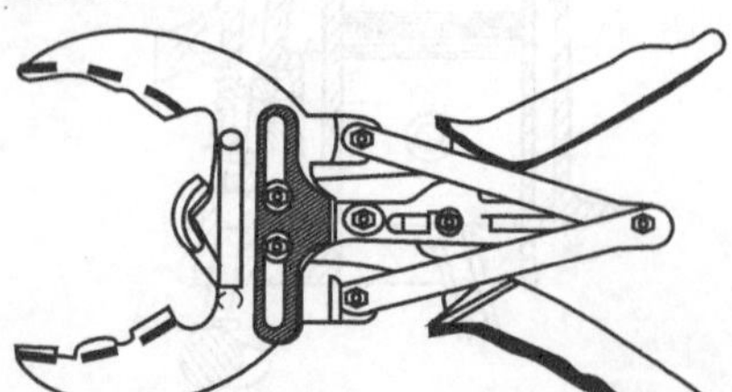

(a) Torque wrench
(b) Piston ring compressor
(c) Piston ring expander
(d) Circlip plier

36. A connecting rod connects the crankshaft and the:
(a) Cylinder head (b) Cylinder block
(c) Piston (d) Camshaft

37. The piston pins in modern automobile engines are usually:
(a) Semi-floating
(b) Fully-floating
(c) Three-quarter floating
(d) Fixed on both the piston and the connecting rod end

38. What is mounted on the rear end of the crankshaft?
(a) Timing sprocket
(b) Flywheel
(c) Vibration damper
(d) Counterweight

39. In a six cylinder engine, the angle between the successive crank throws is:
(a) 60° (b) 90°
(c) 120° (d) 180°

40. What is connected to piston through the piston pin?
(a) Gudgeon pin (b) Connecting rod
(c) Camshaft (d) Rocker arm

41. What joins piston and connecting rod?
(a) Gudgeon pin (b) Rocker arm
(c) Camshaft (d) King pin

42. What transfers energy from piston to crankshaft?
(a) Gudgeon pin (b) King pin
(c) Connecting rod (d) Camshaft

43. Which is the key element in converting reciprocating motion into rotary motion?
(a) Connecting rod (b) Gudgeon pin
(c) King pin (d) Camshaft

44. The opening and closing of valves in relation to piston movement is called:
(a) Valve timing
(b) Valve operation

(c) Valve mechanism
(d) Valve overlap

45. The overlap period is the period during which:
(a) Both inlet and exhaust valves are in closed position
(b) Both inlet and exhaust valves are in open position
(c) Inlet valve is open and exhaust valve is closed
(d) Inlet valve is closed and exhaust valve is open

46. In some engines, the axes of cam lobe and lifter are offset a little. This is done for:
(a) Uniform wear of cam lobe
(b) Uniform wear of lifter
(c) Uniform wear of camshaft's bushes
(d) Uniform opening of valves

47. Which type of camshaft drive mechanism is used when crankshaft and camshaft are very close to each other?
(a) Gear drive (b) Chain drive
(c) Belt drive (d) Sprocket drive

48. What is mounted on the front end of the crankshaft?
(a) Timing gear
(b) Vibration damper
(c) Fan pulley (d) All of these

49. The vibration damper on the crankshaft reduces the:
(a) Longitudinal vibrations
(b) Transverse vibrations
(c) Torsional vibrations
(d) All of these

50. What is the name of the part marked 'X' in the figure given below?

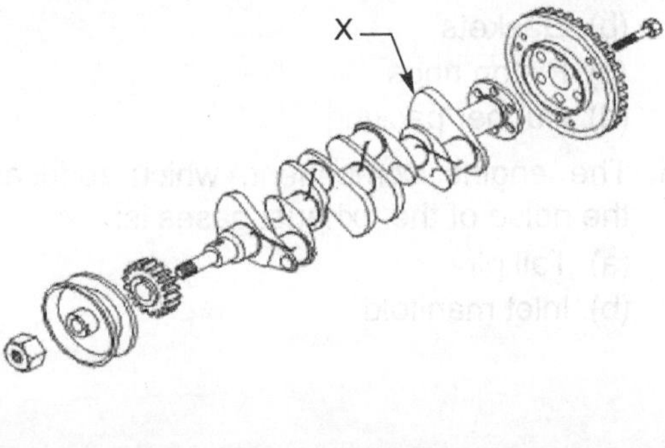

(a) Crank pin
(b) Crank arm
(c) Balancing weight
(d) Main journal

51. What is the name of the part marked 'X' in the figure given below?

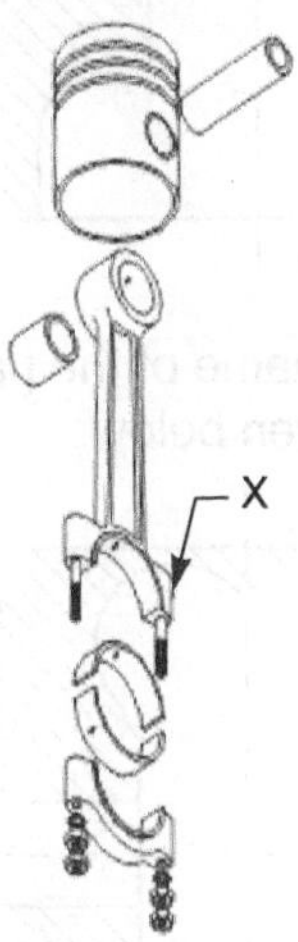

(a) Oil passage (b) Connecting rod
(c) Big end (d) Small end

52. What type of piston is shown in the figure given below?

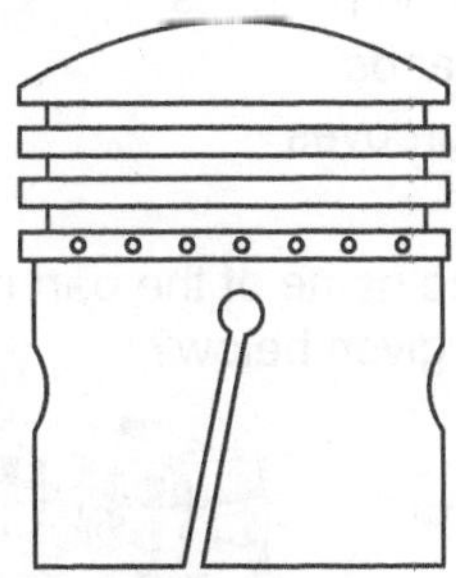

(a) Solid skirt piston
(b) Super piston
(c) Split skirt piston
(d) Piston with steel alloy inserts

53. What is the name of the part marked 'X' in the figure given below?
(a) Gudgeon pin boss
(b) Ring lands
(c) Ring grooves
(d) Crown

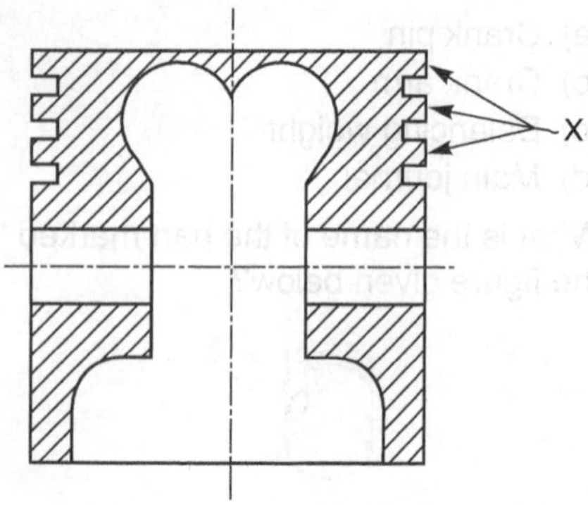

54. What is the name of the part marked 'X' in the figure given below.

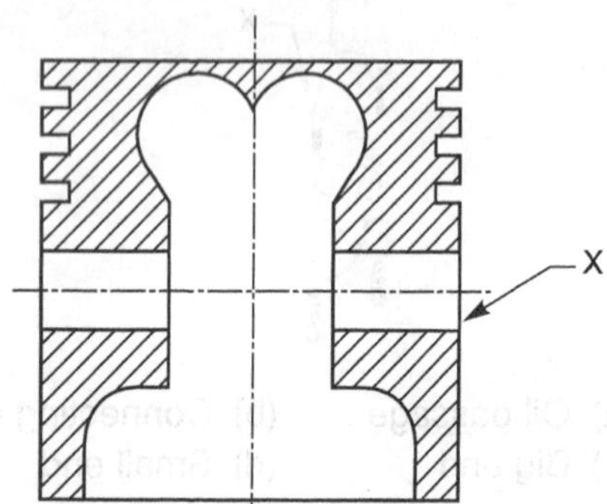

(a) Gudgeon pin boss
(b) Ring lands
(c) Ring grooves
(d) Crown

55. What is the name of the part marked 'X' in the figure given below?

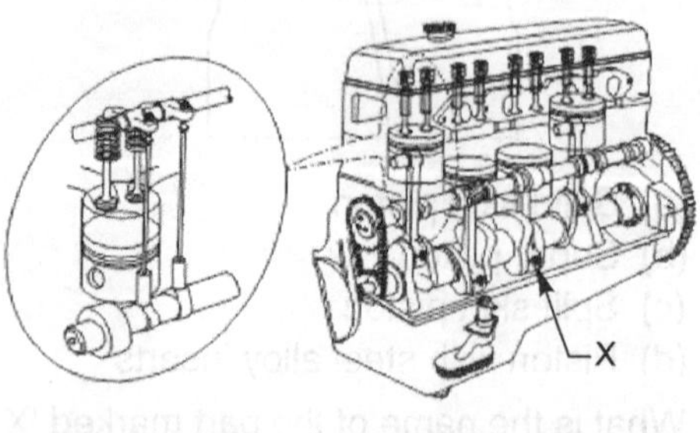

(a) Connecting rod (b) Camshaft
(c) Crankshaft (d) Rocker arm

56. What is the purpose of timing chain?
(a) To connect water pump pulley
(b) To connect alternator
(c) To connect crankshaft to camshaft
(d) To connect A/C compressor

57. What is the purpose of flywheel timing mark?
(a) To make the gears coincide
(b) To set the engine timing
(c) To set the F.I.P. timing
(d) To set the valve clearance

58. What is the effect of taper and ovality of the engine cylinder bore?
(a) Compression loss
(b) Misfiring
(c) Difficult starting
(d) Wrong valve timing

59. What type of bearing is fitted in the connecting rod big end?
(a) Needle bearing
(b) Ball bearing
(c) Taper roller bearing
(d) Shell bearing

60. What is the term used to refer when a valve opens before TDC?
(a) Lead (b) Lag
(c) Overlap (d) Retard

61. What is the term used to refer when a valve closes after BDC?
(a) Lead (b) Advance
(c) Lag (d) Overlap

62. What is the term used to refer when both valves are in open position?
(a) Lead (b) Lag
(c) Overlap (d) Advance

63. An engine valves are closed by:
(a) Crankshaft (b) Camshaft
(c) Timing device (d) Valve springs

64. Which of the following are added to the piston to ensure good seal between piston and cylinder?
(a) Piston pins
(b) Gaskets
(c) Piston rings
(d) Rubber packing

65. The engine component, which reduces the noise of the exhaust gases is:
(a) Tail pipe
(b) Inlet manifold

(c) Muffler
(d) Exhaust manifold

66. In which one of the following types the valves are arranged in the cylinder head?
(a) I – head (b) L – head
(c) F – head (d) T – head

67. In which one of the following types the inlet valve is arranged in the cylinder head and the exhaust valve in the cylinder block?
(a) I – head (b) L – head
(c) F – head (d) T – head

68. In multi-cylinder engine, which one of the following arrangements is considered to be the simplest?
(a) Cylinders in – line
(b) V – engines
(c) Horizontally opposed type
(d) Radial

69. For engines having eight to sixteen cylinders, which one of the following types of cylinder arrangement is considered to be appropriate?
(a) Cylinders in – line
(b) V – engines
(c) Horizontally opposed type
(d) Radial

ANSWERS

1. (a)	**2.** (a)	**3.** (b)	**4.** (c)	**5.** (b)	**6.** (c)	**7.** (a)	**8.** (a)	**9.** (d)	**10.** (b)
11. (c)	**12.** (a)	**13.** (d)	**14.** (d)	**15.** (c)	**16.** (a)	**17.** (c)	**18.** (a)	**19.** (a)	**20.** (d)
21. (a)	**22.** (a)	**23.** (d)	**24.** (d)	**25.** (b)	**26.** (c)	**27.** (d)	**28.** (c)	**29.** (d)	**30.** (a)
31. (b)	**32.** (b)	**33.** (a)	**34.** (a)	**35.** (c)	**36.** (c)	**37.** (b)	**38.** (b)	**39.** (c)	**40.** (b)
41. (a)	**42.** (c)	**43.** (a)	**44.** (a)	**45.** (b)	**46.** (b)	**47.** (a)	**48.** (d)	**49.** (c)	**50.** (c)
51. (c)	**52.** (c)	**53.** (b)	**54.** (a)	**55.** (c)	**56.** (c)	**57.** (b)	**58.** (a)	**59.** (d)	**60.** (a)
61. (c)	**62.** (c)	**63.** (d)	**64.** (c)	**65.** (c)	**66.** (a)	**67.** (c)	**68.** (a)	**69.** (b)	

EXPLANATORY NOTES

5: Main parts of a crankshaft are shown in the figure given below.

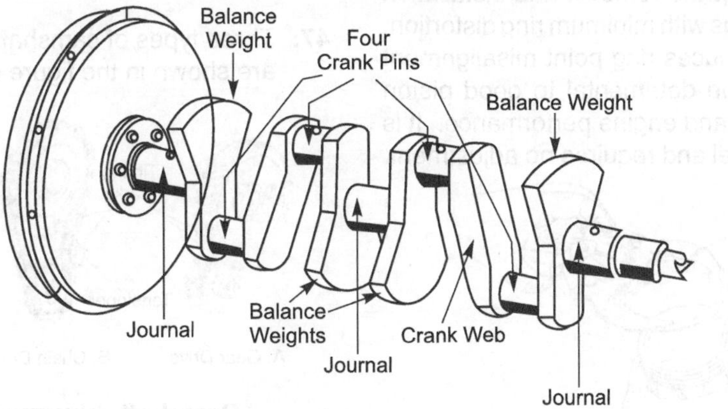

Parts of crankshaft

7: Piston and its sections and parts are shown in the figure given below.

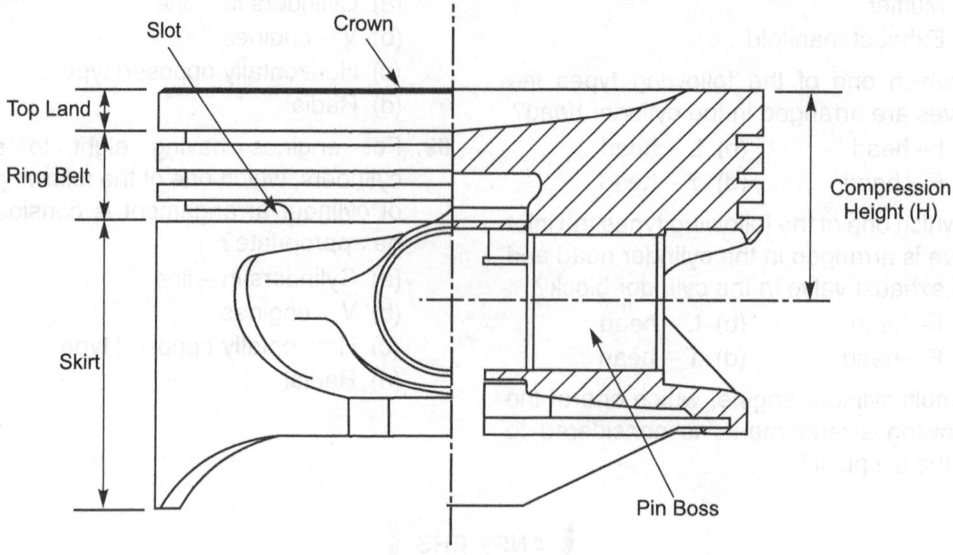

Piston

18: In internal combustion engines, the gudgeon pin connects the piston to the connecting rod, and provides a bearing for the connecting rod to pivot upon as the piston moves.

32: A tappet is also known as cam follower or valve lifter. It follows the shape of the cam to provide reciprocating motion to the valves by converting angular/rotating motion of the cam shaft.

35: A piston ring expander is a unique tool that will speed up the removal and installation of piston rings with minimum ring distortion. It greatly reduces ring point misalignment which can be detrimental to good piston ring sealing and engine performance. It is made of steel and requires no adjustment.

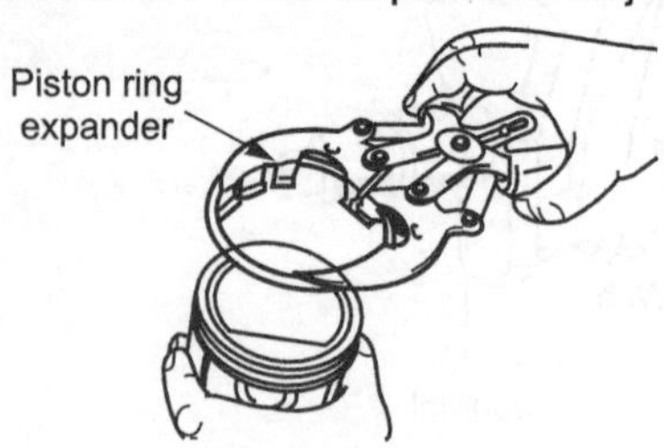

Piston ring expander

39: A 6-cylinder engine is simply a pair of 3-cylinder engines mated together symmetrically, thus piston 1 is always in the same position as piston 6, piston 2 the same as piston 5. In other words, the engine is balanced end-to-end and requires no balancer shaft, unlike 3-cylinder engines.

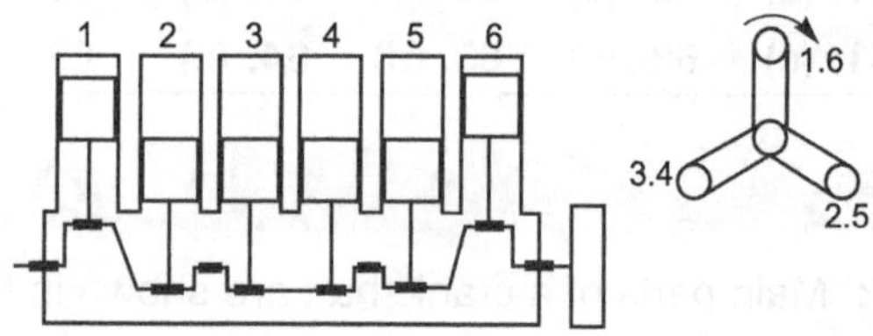

47: Three types of camshaft drive mechanism are shown in the figure given below.

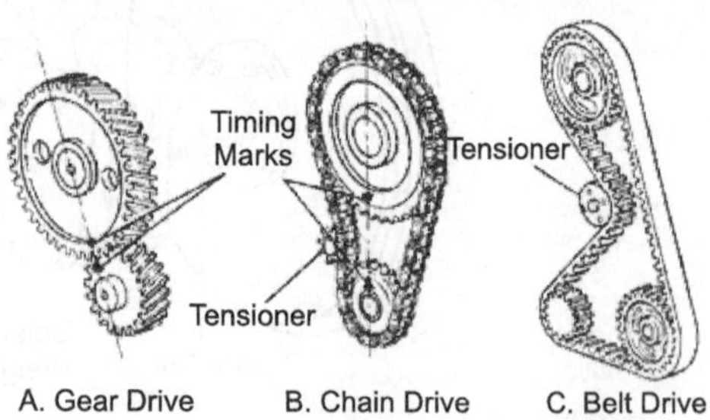

Camshaft drive mechanisms

52: As a split-skirtpiston warms up, the split merely closes, thereby keeping the skirt from expanding outward and binding the piston in the cylinder.

Module 10

Cooling and Lubrication System

MULTIPLE CHOICE QUESTIONS

1. Which type of cooling system uses fins on the cylinder head?
(a) Oil cooling (b) Water cooling
(c) Liquid cooling (d) Air cooling

2. Which part helps in dissipating heat in an air-cooled engine?
(a) Engine piston
(b) Engine crankshaft
(c) Density of water
(d) Cylinder and head fins

3. In which engine cooling system, the rate of cooling is very low?
(a) Thermo-siphon system
(b) Air cooling system
(c) Forced feed system
(d) Pump circulation system

4. In forced feed engine cooling system, which part forces water to circulate?
(a) Pump (b) Thermostat
(c) Density of water (d) Water jacket

5. What is the name of the part of marked 'X' in the figure given below?

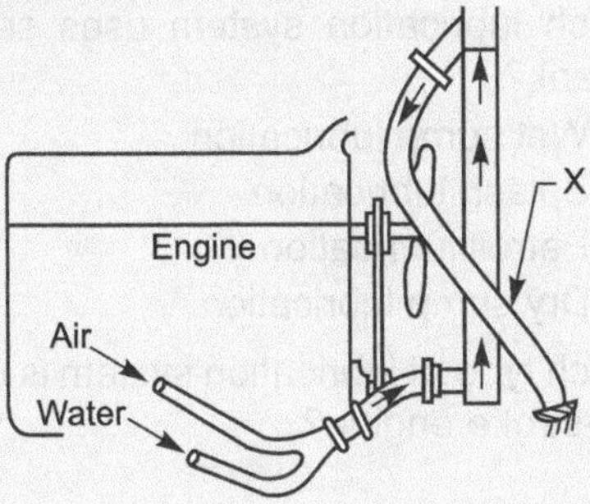

(a) Air hose
(b) Radiator drain hose
(c) Radiator inlet
(d) Fuel hose

6. What is the name of the part of marked 'X' in the figure given below?

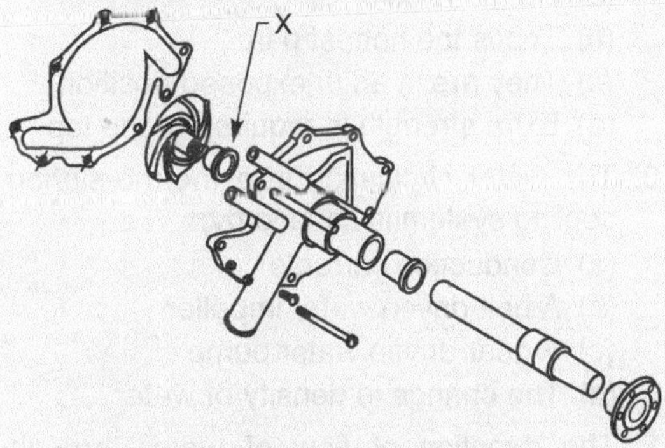

(a) Bearing (b) Shaft
(c) Oil seal (d) Impeller

7. What is the name of the part of marked 'X' in the figure given below?

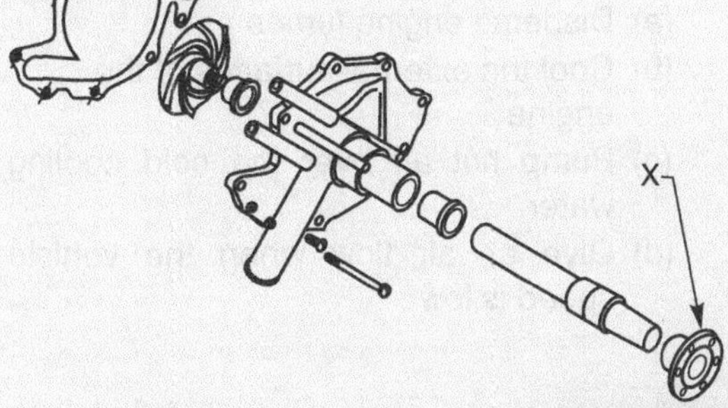

(a) Shaft (b) Impeller
(c) Pulley hub (d) Housing

8. What is the name of the part of marked 'X' in the figure given below?

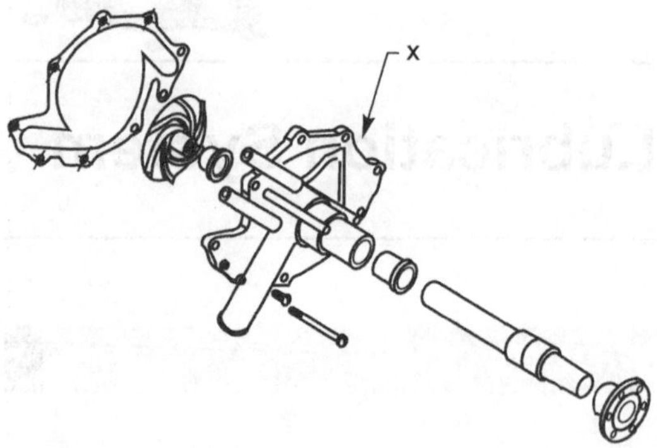

(a) Impeller (b) Pulley
(c) Housing (d) Shaft

9. The fins at the top of a motor cycle engine cylinder are longer than those at the bottom, because:
(a) Hot air rises
(b) Top is the hottest part
(c) They are in an unexposed position
(d) Extra strength is required at the top

10. The water circulation in a thermo-siphon cooling system is caused by:
(a) Conduction currents
(b) A belt driven water impeller
(c) A gear driven water pump
(d) The change in density of water

11. The direction of flow of water through the radiator of a thermo-siphon cooling system is from:
(a) Top to bottom (b) Bottom to top
(c) Front to back (d) Back to front

12. The main purpose of the fan of a liquid cooling system for an engine is to:
(a) Disperse engine fumes
(b) Cool the external surfaces of the engine
(c) Pump hot air over the cold cooling water
(d) Give an air flow when the vehicle speed is low

13. The principle of a radiator of an engine cooling system is to:
(a) Act as a reservoir for water
(b) Cause heat flow by convection currents
(c) Spread out hot water over a large area
(d) Increase air speed as it flows over hot surface

14. What is the name of the part marked 'X' in the figure given below?

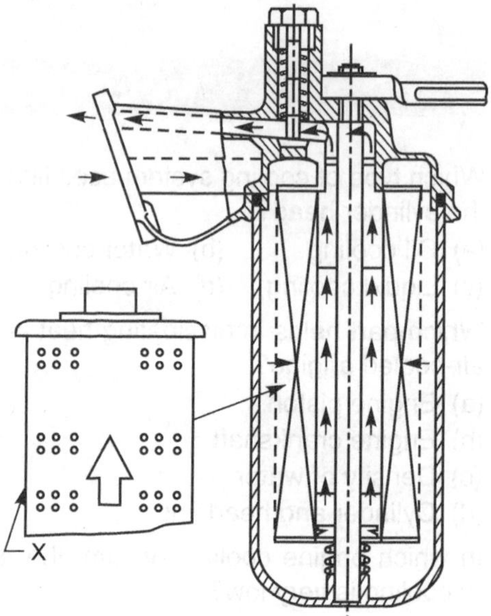

(a) Air filter (b) Oil filter
(c) Water filter (d) Fuel filter

15. What is a desirable property of a lubricant?
(a) Boiling temperature should be low
(b) Should develop foam
(c) Viscosity should not be the same in hot and cold condition
(d) Viscosity should suit the operating conditions

16. Which lubrication system uses separate oil tank?
(a) Wet sump lubrication
(b) Splash lubrication
(c) Petroil lubrication
(d) Dry sump lubrication

17. Which type of lubrication system is used in two-stroke engine?

(a) Dry sump lubrication
(b) Wet sump lubrication
(c) Petroil lubrication
(d) Splash lubrication

18. Which part of engine is marked 'X' in the figure given below?

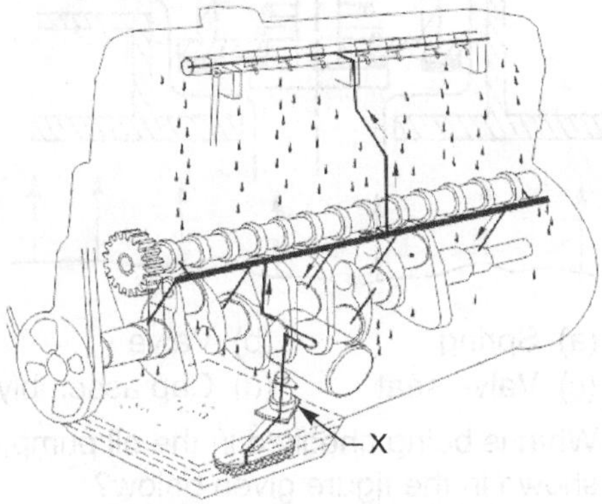

(a) Suction pump (b) Oil filter
(c) Oil pump (d) Oil strainer

19. Which part of the engine drives the oil pump of lubrication system?
(a) Crankshaft (b) Camshaft
(c) Crank pulley (d) Timing gears

20. Which device sucks oil from the oil sump?
(a) Strainer (b) Suction pipe
(c) Pump (d) Filter

21. Under which condition does thermostat valve of engine cooling system open?
(a) At low temperature of the engine
(b) At high temperature of the engine
(c) At operating temperature of the engine
(d) At freezing temperature of the engine

22. What cleaning method is shown in the figure given below?

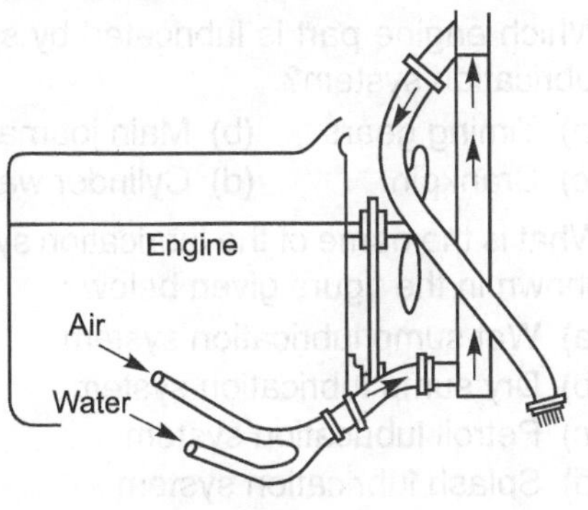

(a) Pressure flushing system
(b) Reverse flushing system
(c) Engine cooling system
(d) Water draining system

23. Which method is used in radiator reverse flushing cleaning?
(a) Flushing water with air (gun) pressure
(b) Flushing water with engine oil
(c) Flushing water with coolant oil
(d) Flushing water with soap oil

24. Where the oil cooler is fitted in an engine?
(a) Engine block
(b) Cylinder head
(c) Exhaust manifold
(d) Oil sump

25. Which method used to de-scale the water passages in the engine block?
(a) By injecting water solvent
(b) By injecting oil solvent
(c) By injecting fresh air
(d) By cleaning tools

26. How are damaged radiator core tubes checked?
(a) Visually
(b) With special tools
(c) With computer
(d) With bore dial gauge

27. What is the name of testing marked 'X' in the figure given below?

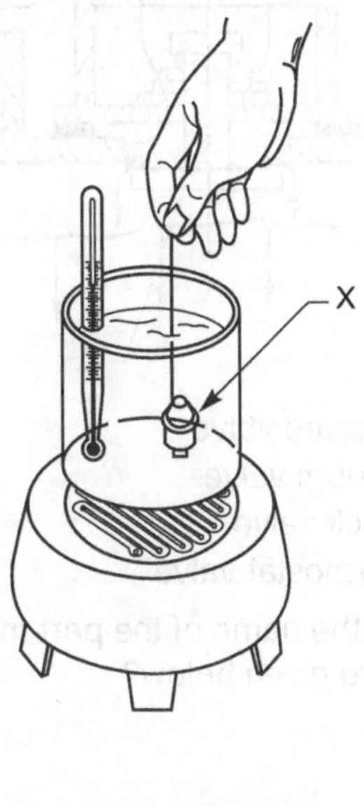

(a) Thermometer testing
(b) Thermostat valve testing
(c) Water pump testing
(d) Impeller testing

28. What is the name of the part marked 'X' in the figure given below?

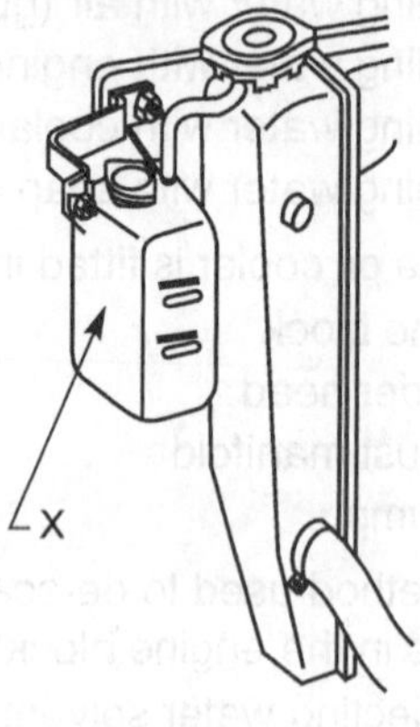

(a) Expansion tank
(b) Upper tank
(c) Lower tank
(d) Oil tank

29. What is the name of the part marked 'X' in the figure given below?

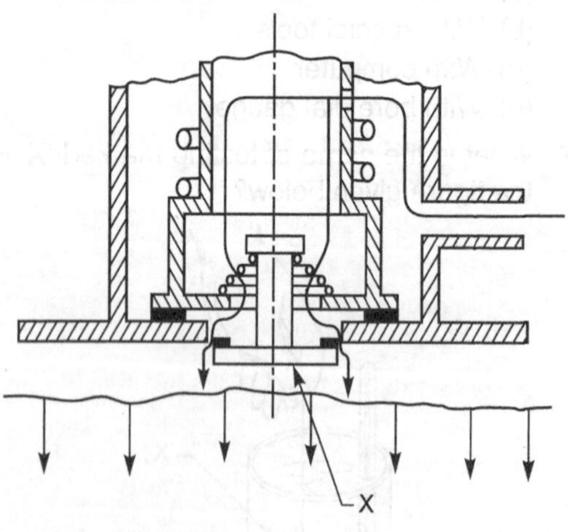

(a) Pressure valve
(b) Vacuum valve
(c) Check valve
(d) Thermostat valve

30. What is the name of the part marked 'X' in the figure given below?

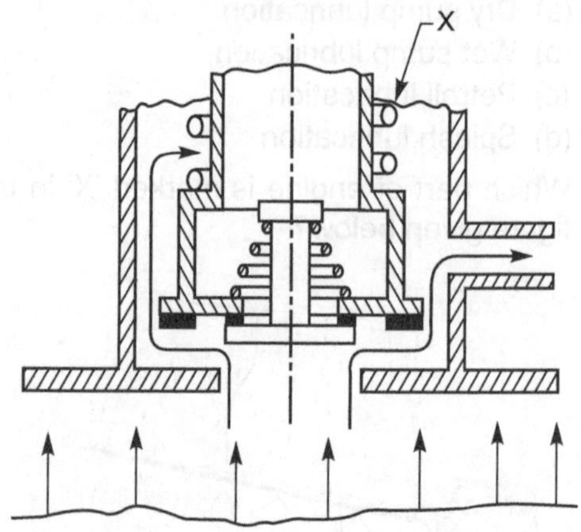

(a) Spring (b) Valve
(c) Valve seat (d) Cap assembly

31. What is being checked in the oil pump, as shown in the figure given below?

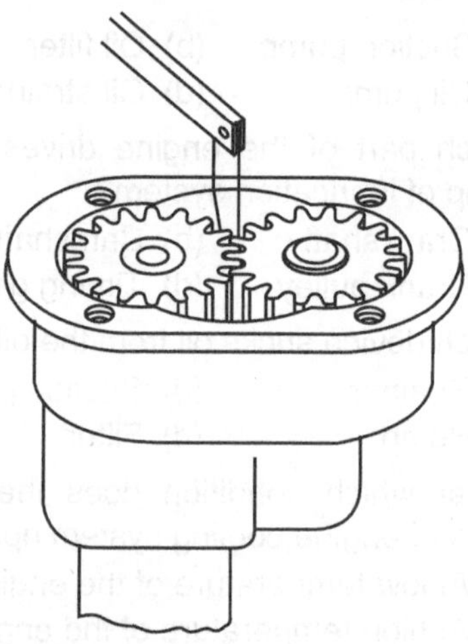

(a) Radial clearance
(b) Backlash clearance
(c) Surface
(d) Depth

32. Which engine part is lubricated by splash lubrication system?

(a) Timing gears (b) Main journal
(c) Crankpin (d) Cylinder walls

33. What is the name of the lubrication system shown in the figure given below?

(a) Wet sump lubrication system
(b) Dry sump lubrication system
(c) Petroil lubrication system
(d) Splash lubrication system

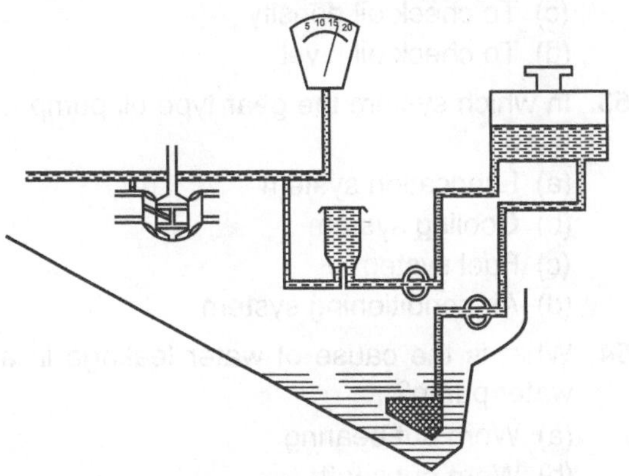

34. How oil is delivered from the crankshaft main bearings to the connecting rod bearings?
(a) Through drilled oil passages
(b) Through crank pulley
(c) Through sprocket
(d) Through vibration damper

35. What is the name of the part marked 'X' of two-stroke engine in the figure given below?

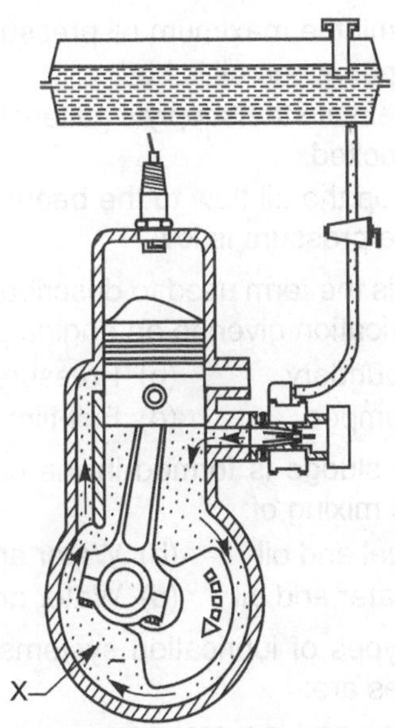

(a) Oil tank (b) Wet oil sump
(c) Dry oil sump (d) Crankcase

36. What is the purpose of water pump in the engine cooling system?
(a) To force water circulation
(b) To reduce pressure of water
(c) To increase temperature of water
(d) To flush out the cooling system

37. What will happen if thermostat valve is stuck?
(a) Engine will get over-cooled
(b) Engine will get overheated
(c) Engine will not start
(d) Engine will not stop

38. Where does the dissipation of heat take place in engine cooling system?
(a) Radiator (b) Water pump
(c) Water jackets (d) Fan

39. Which type of pump is used in water cooling system?
(a) Gear pump
(b) Diaphragm pump
(c) Centrifugal pump
(d) Reciprocating pump

40. It is necessary to be very careful when removing the pressure cap when the engine is hot. It is so, because:
(a) The seal may be damaged
(b) Of the risk of scalding
(c) The cooling system pressure is lower than atmospheric pressure
(d) The sudden increase in pressure can damage the radiator

41. A large spring loaded valve is provided in the radiator cap. What is its purpose?
(a) It lowers the temperature at which coolant boils
(b) It prevents the coolant escaping when it boils
(c) It reduces the risk of the rubber hoses collapsing when the pressure is low
(d) It pressurises the system which raises the boiling point of the coolant

42. Which one of the following types of thermostats is most commonly used in modern engines?
(a) Wax-pellet type (b) Bi-metal type
(c) Bellows type (d) Alcohol type

43. In order to ensure effective cooling of each cylinder of the engine is surrounded by:

(a) Radiator (b) Water tube
(c) Water jacket (d) Water ring

44. One of the components of the pump makes water circulate in the cooling system. This component is:
(a) Fan (b) Body
(c) Impeller (d) Pulley

45. On leaving the engine, the coolant goes to:
(a) Pump inlet (b) Header tank
(c) Collector tank (d) Jacket

46. The radiator core is made of:
(a) Plastic (b) Cast iron
(c) Steel (d) Brass

47. A pressure cap contains:
(a) Pressure valve
(b) Thermostat valve
(c) Blow-off valve
(d) Pressure and vacuum valve

48. What is likely to happen if the thermostat valve remains in open position?
(a) Slow warming of the engine
(b) Overheating of the engine
(c) Difficulty in starting of the engine
(d) Stalling of the engine

49. The boiling point of the coolant is increased in the cooling system. Which part makes it possible?
(a) Water jackets
(b) Vacuum valve
(c) Pressure radiator cap
(d) Radiator core

50. Which part allows water to flow from upper tank to the lower tank of the radiator?
(a) Fins (b) Core tubes
(c) Water pump (d) Bottom hole

51. What is the purpose of metal fins in air cooling system?
(a) To supply heat (b) To increase heat
(c) To reduce heat (d) To maintain heat

52. What is the purpose of using dipstick in an engine?
(a) To check oil pressure
(b) To check oil temperature
(c) To check oil density
(d) To check oil level

53. In which system the gear type oil pump is used?
(a) Lubrication system
(b) Cooling system
(c) Fuel system
(d) Air conditioning system

54. What is the cause of water leakage in a water pump?
(a) Worn out bearing
(b) Worn out shaft
(c) Worn out seal
(d) Worn out impeller

55. What is the cause of pump not sucking oil?
(a) Less radial clearance
(b) More backlash
(c) Relief valve stuck
(d) Filter clogged

56. A relief valve is fitted to the main gallery of an engine. The purpose of this valve is to:
(a) Limit the maximum oil pressure
(b) Open when oil is hot
(c) Maintain the supply if gallery becomes blocked
(d) Stop the oil flow to the bearings when the pressure is low

57. What is the term used to describe the type of lubrication given to an engine piston?
(a) Boundary (b) Pressure
(c) Pumped (d) Full-film

58. Water sludge is formed in the crankcase due to mixing of:
(a) Fuel and oil (b) Water and fuel
(c) Water and oil (d) Water and air

59. Two types of lubrication systems used in engines are:
(a) High and low pressure
(b) Pressure and vacuum
(c) Pump and gravity
(d) Splash and pressure

60. The contamination of lubricating oil in an engine takes place due to:

(a) Worn out piston and piston rings
(b) Worn out connecting rod bearings
(c) Loose gudgeon pin
(d) High pressure in lubrication system

61. Among the following lubricating oils, which one has the highest viscosity?
(a) SAE 30 (b) SAE 40
(c) SAE 50 (d) SAE 80

ANSWERS

1. (d)	**2.** (d)	**3.** (a)	**4.** (a)	**5.** (b)	**6.** (c)	**7.** (c)	**8.** (c)	**9.** (b)	**10.** (d)
11. (a)	**12.** (d)	**13.** (c)	**14.** (b)	**15.** (d)	**16.** (d)	**17.** (c)	**18.** (c)	**19.** (b)	**20.** (c)
21. (c)	**22.** (b)	**23.** (a)	**24.** (a)	**25.** (a)	**26.** (a)	**27.** (b)	**28.** (a)	**29.** (b)	**30.** (a)
31. (b)	**32.** (d)	**33.** (b)	**34.** (a)	**35.** (d)	**36.** (a)	**37.** (b)	**38.** (a)	**39.** (c)	**40.** (b)
41. (d)	**42.** (a)	**43.** (c)	**44.** (c)	**45.** (b)	**46.** (d)	**47.** (d)	**48.** (a)	**49.** (c)	**50.** (b)
51. (c)	**52.** (d)	**53.** (a)	**54.** (c)	**55.** (b)	**56.** (a)	**57.** (a)	**58.** (c)	**59.** (d)	**60.** (a)
61. (d)									

EXPLANATORY NOTES

3: In thermo-siphon cooling system, cooling water becomes less dense as it is heated. It moves up in convection current to the radiator header tank. It displaces colder and denser water from the radiator bottom tank. Passing downward through the radiator core, heat is transferred to air current, created by the movement of the vehicle generally assisted by a fan.

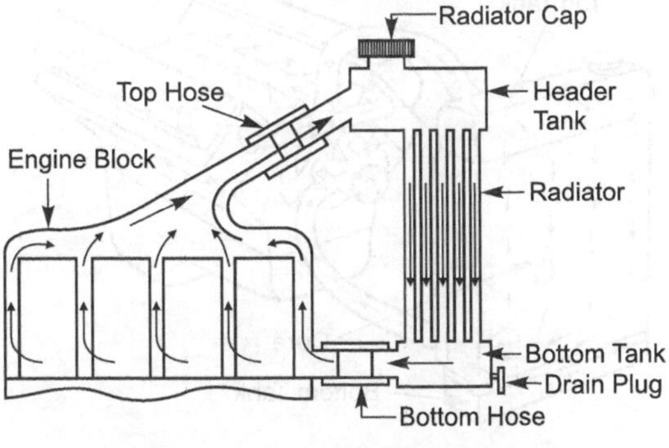

Thermo-siphon cooling system

16: In dry sump system, the oil is carried in a separate tank from where it is fed to the engine. The oil falls into the sump after lubrication. It is then pumped back to the oil tank by a separate delivery pump. There are, therefore, two pumps in this system.

17: Petroil Lubrication: This system of lubrication is generally adopted in two stroke petrol engines, like scooters and motorcycles. The lubricating oil is mixed into the petrol itself while filling in the petrol tank of the vehicle, in a specified ratio.

18: Pressure System: In this lubrication system engine parts are lubricated under pressure feed. The lubricating oil is stored in a separate tank or the sump, from which an oil pump receives the oil through a strainer and transfers it through a filter to the central oil gallery at a pressure of 2-4 kg/cm^2.

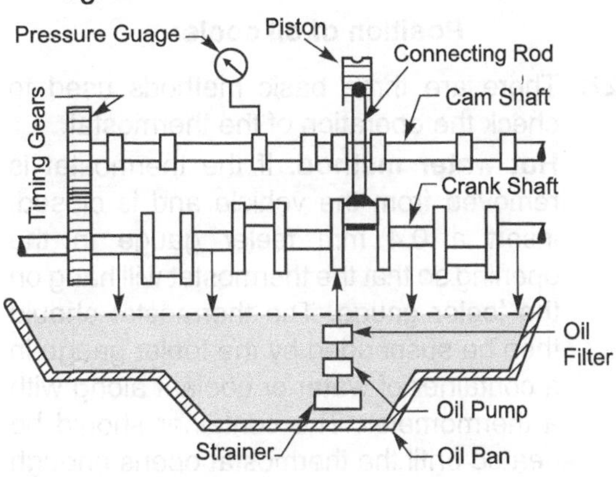

Pressure system of lubrication

21: An engine works very poorly when it is cold. It uses a lot of fuel, produces less power, and is subject to much more wear than when it is running at normal temperature. It is, therefore, important to allow the engine to warm up to its normal running temperature as quickly as possible, which means 'turning off' the cooling system. The thermostat serves this purpose. The thermostat is a valve which opens automatically at the pre-determined temperature. It is placed in the water pipe leading from the engine to the radiator.

22: Reverse flushing involves flowing air and water backwards to the normal flow of coolant through the radiator core and engine block.

24: The position of oil cooler is shown in the figure given below.

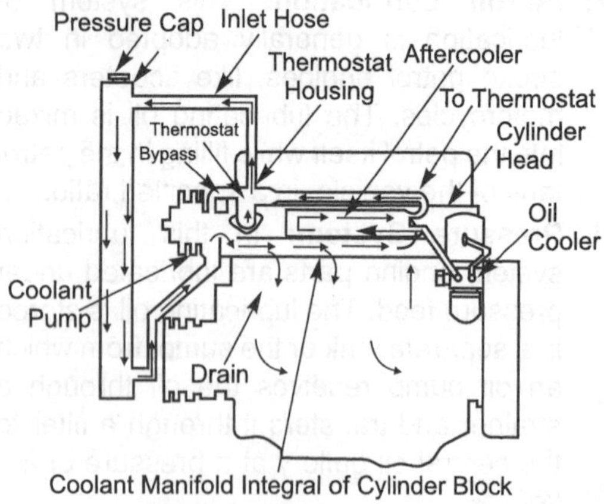

Position of oil cooler

27: There are three basic methods used to check the operation of the thermostat.

Hot water method. If the thermostat is removed from the vehicle and is closed, insert a 0.4 mm feeler gauge in the opening so that the thermostat will hang on the feeler gauge. The thermostat should then be suspended by the feeler gauge in a container of water or coolant along with a thermometer. The container should be heated until the thermostat opens enough to release and fall from the feeler gauge. The temperature at which the thermostat falls is the opening temperature of the thermostat. If it is within 4°C of the temperature stamped on the thermostat, the thermostat is satisfactory for use. If the temperature difference is greater, the thermostat should be replaced.

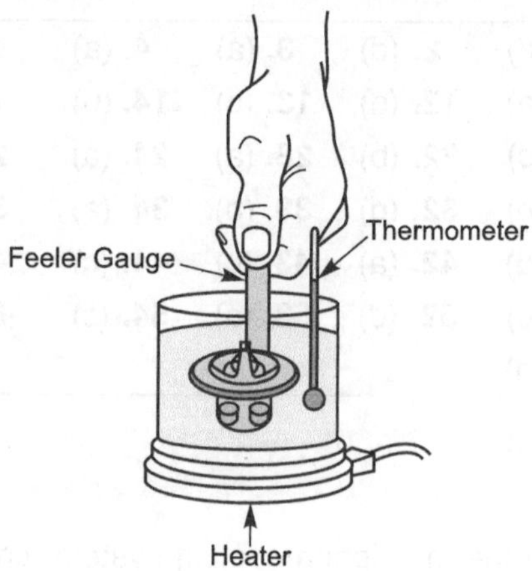

Testing of thermostat

The other two methods are Infrared thermometer method and Scan tool method.

50: The figure given below shows how water flows from top to bottom tank of a radiator.

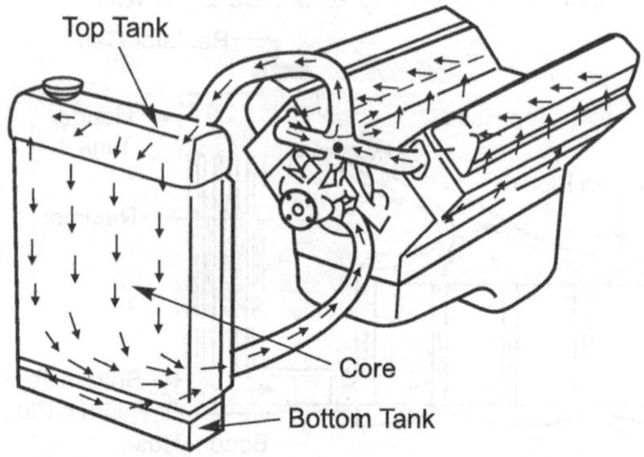

Water flow from top to bottom tank of radiator

61: Oils are classified by the SAE number – the higher the number the thicker the oil is, e.g. an SAE 40 oil is thicker than SAE 30 oil.

Module 11

Intake and Exhaust System

MULTIPLE CHOICE QUESTIONS

1. Which part of petrol engine is connected to inlet manifold?
(a) Air filter (b) Injector
(c) Carburettor (d) Fuel pump

2. What material inlet manifold is made of?
(a) Brass
(b) Bronze
(c) Stainless steel
(d) Aluminium alloy

3. Which component is related to exhaust system of an engine?
(a) Muffler (b) Air filter
(c) Carburettor (d) Injection pump

4. Where is a turbocharger mounted in an engine?
(a) At air filter
(b) At inlet manifold
(c) At cylinder head
(d) At exhaust manifold

5. In a turbocharged engine, the air pressure is increased by the turbine – compressor, which is driven by:
(a) Crankshaft (b) Fan pulley
(c) Starter motor (d) Exhaust gases

6. In a diesel engine, what is drawn in the cylinder during suction stroke?
(a) Gas
(b) Fuel alone
(c) Pure air alone
(d) Mixture of air and fuel

7. Name the type of muffler shown in the figure given below.

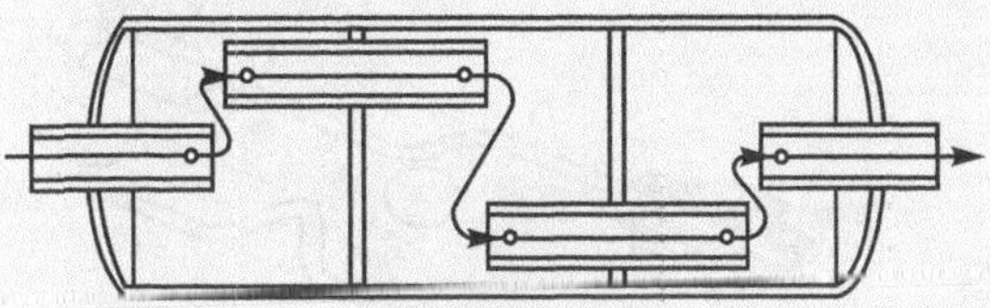

(a) Reverse flow muffler
(b) Straight through muffler
(c) Baffle type muffler
(d) Electronic muffler

8. Name the type of muffler shown in the figure given below.

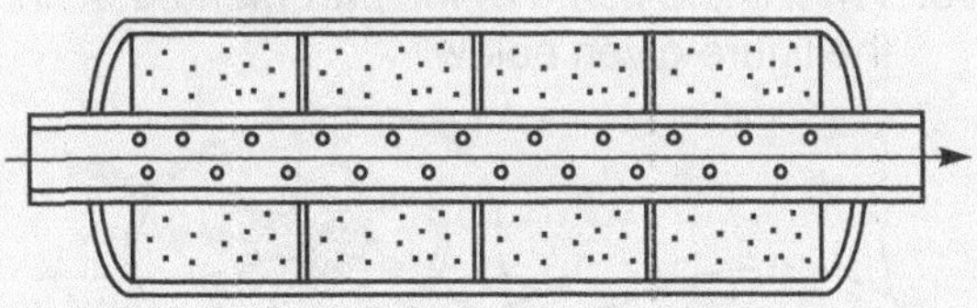

(a) Reverse flow muffler
(b) Straight through muffler
(c) Baffle type muffler
(d) Electronic muffler

9. Name the type of muffler shown in the figure given below.

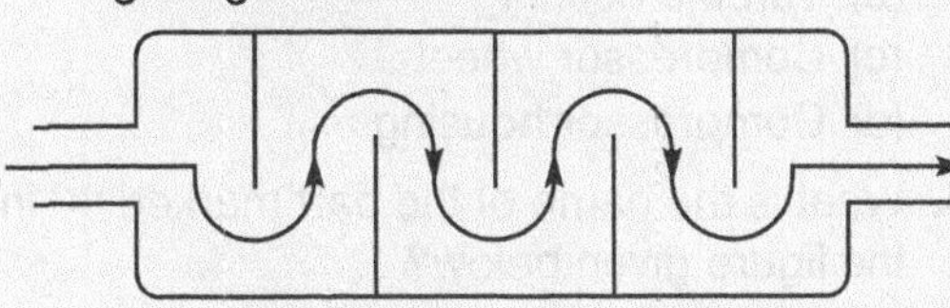

(a) Reverse flow muffler

(b) Straight through muffler
(c) Baffle type muffler
(d) Electronic muffler

10. Which type of muffler is fitted with sensors and microphone?
(a) Electronic type
(b) Resonance type
(c) Reverse flow type
(d) Straight through type

11. The process which takes place in a catalytic converter is:
(a) Thermal (b) Physical
(c) Chemical (d) Biological

12. What is the name of the part marked 'X' in the figure given below?

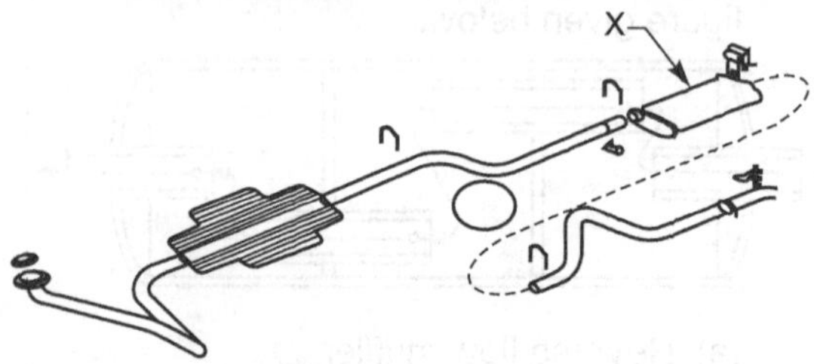

(a) Muffler
(b) Tailpipe
(c) Exhaust manifold
(d) Catalytic converter

13. What is the name of the part marked 'X' in the figure given below?

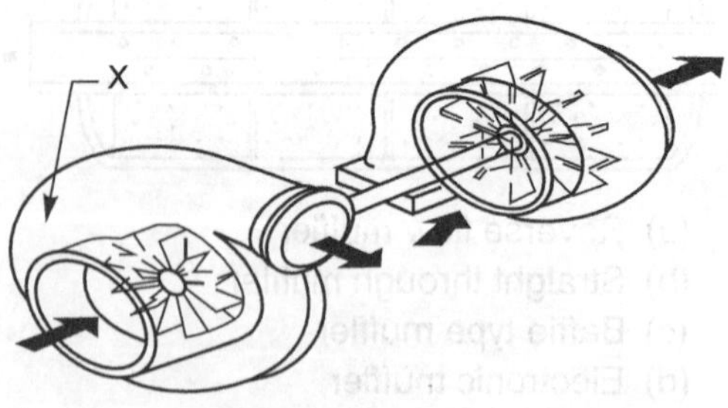

(a) Turbine wheel
(b) Turbine housing
(c) Compressor wheel
(d) Compressor housing

14. What is the name of the part marked 'X' in the figure given below?

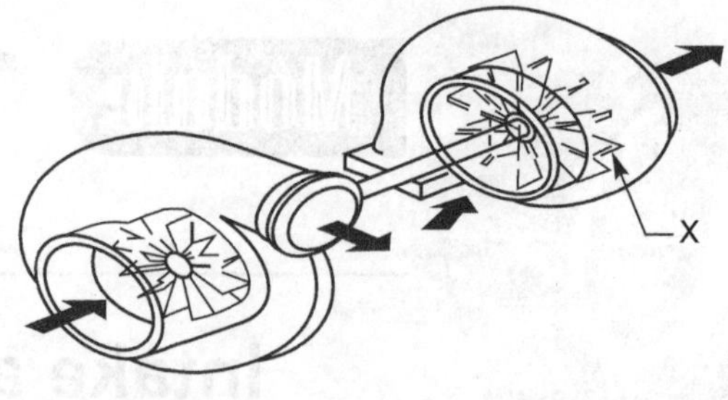

(a) Turbine wheel
(b) Turbine housing
(c) Compressor wheel
(d) Compressor housing

15. What is the purpose of a turbocharger in an engine?
(a) To save power
(b) To do scavenging
(c) To filter exhaust gases
(d) To pump more air into the cylinder

16. Which one of the following acts as a flame arrester during engine backfire?
(a) Muffler (b) Oil filter
(c) Air filter (d) Fuel filter

17. Which type of filter element is fitted in dry type air cleaner?
(a) Cloth (b) Paper
(c) Strainer (d) Wire mesh

18. What is the function of the muffler in the engine exhaust system?
(a) To reduce heat
(b) To reduce noise
(c) To reduce vibrations
(d) To filter exhaust gases

19. Supercharging in a diesel engine is also called:
(a) Scavenging (b) Boosting
(c) Cleaning (d) Silencing

20. What is the effect of carbon deposit in the exhaust system of an engine?
(a) Increase in back pressure
(b) Decrease in back pressure
(c) No effect on back pressure
(d) Fluctuation in back pressure

21. What is purpose of scavenging in a diesel engine?

(a) To maintain fuel consumption
(b) To maintain power output
(c) To maintain speed
(d) To maintain oil pressure

22. Which of these causes reduction of exhaust noise?
(a) Muffler
(b) Exhaust pipe
(c) Exhaust manifold
(d) Tail pipe

23. In engine exhaust system, muffler is connected between:
(a) Exhaust pipe and tailpipe
(b) Tailpipe and exhaust manifold
(c) Engine head and exhaust manifold
(d) Exhaust pipe and exhaust manifold

24. Catalytic converter is used in:
(a) Fuel system
(b) Inlet system
(c) Exhaust system
(d) Lubrication system

25. What is the function of catalytic converter?
(a) To control noise
(b) To control emission
(c) To control temperature
(d) To control fuel consumption

26. A catalytic converter is fitted between:
(a) Muffler and tailpipe
(b) Exhaust pipe and muffler
(c) Tailpipe and resonator pipe
(d) Engine head and exhaust manifold

27. What causes decrease in volumetric efficiency of exhaust system?
(a) Low noise
(b) High temperature
(c) Low temperature
(d) Excessive back pressure

28. Which metal is to be avoided while fitting catalytic converter?
(a) Tin (b) Lead
(c) Chromium (d) Phosphorous

29. Two types of scavenging processes are used to supply fresh air into the cylinder. In loop scavenging, the top of the piston is:
(a) Flat shaped
(b) Contoured shaped
(c) Slanted shaped
(d) Concave shaped

30. What normally drives the turbocharger in a diesel engine?
(a) Incoming air (b) Exhaust gases
(c) Lub. oil pump (d) Camshaft

31. If the temperature of the intake air in an internal combustion engine increases, what will be its effect on engine efficiency?
(a) Engine efficiency will increase
(b) Engine efficiency will decrease
(c) Engine efficiency will remain the same
(d) It is difficult to predict the effect on engine efficiency

32. The operation of forcing additional air under pressure in the engine cylinder is known as:
(a) Scavenging (b) Supercharging
(c) Turbulence (d) Hammering

33. The size of inlet valve in comparison to exhaust valve is:
(a) More
(b) Less
(c) Same
(d) More or less depending upon capacity of the engine

34. The output of a diesel engine can be increased by:
(a) Feeding more fuel
(b) Heating incoming air
(c) Scavenging
(d) Supercharging

35. The colour of exhaust gases from a diesel engine is generally:
(a) White (b) Bluish
(c) Black (d) Violet

36. A smoky blue exhaust from an engine is due to:
(a) Lean mixture
(b) Excessive fuel consumption
(c) Burning of oil in the combustion chamber
(d) Incorrect valve adjustment

37. Air flowing out of the engine crankcase is drawn back into the inlet manifold and fed into the engine. This is known as:
(a) Positive crankcase ventilation
(b) Negative crankcase ventilation
(c) Zero crankcase ventilation
(d) Open type crankcase ventilation

38. SCR stands for:
(e) Selective Catalytic Reduction
(f) Selective Catalytic Regulator
(g) Silicon Catalytic Regulator
(h) Silicon Catalytic Recirculation

ANSWERS

1. (c)	**2.** (d)	**3.** (a)	**4.** (d)	**5.** (d)	**6.** (c)	**7.** (a)	**8.** (b)	**9.** (c)	**10.** (a)
11. (c)	**12.** (a)	**13.** (d)	**14.** (a)	**15.** (d)	**16.** (c)	**17.** (b)	**18.** (b)	**19.** (b)	**20.** (a)
21. (b)	**22.** (a)	**23.** (a)	**24.** (c)	**25.** (b)	**26.** (b)	**27.** (d)	**28.** (b)	**29.** (b)	**30.** (b)
31. (b)	**32.** (b)	**33.** (a)	**34.** (d)	**35.** (c)	**36.** (c)	**37.** (a)	**38.** (a)		

EXPLANATORY NOTES

13: A turbocharger is mounted on the exhaust manifold. It has a turbine wheel and a compressor wheel on the same shaft.

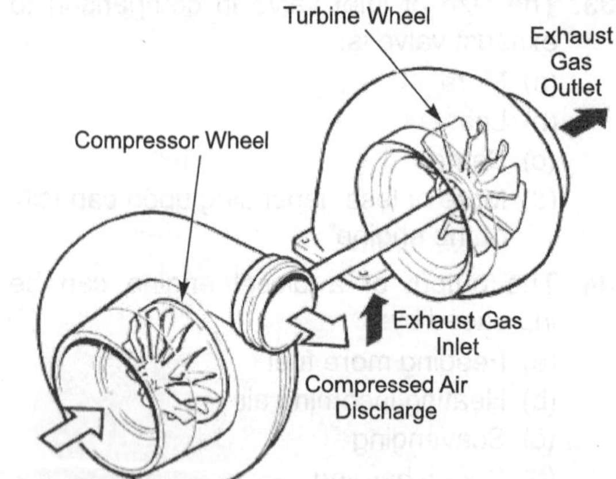

Turbocharger

17: Unlike normal paper used for writing, a special filter with small pores that looks like paper is used as an air filter which provides filtering of air by restricted the flow of solid contaminants and dust particles.

21: Two methods of scavenging are shown in the figure given below.

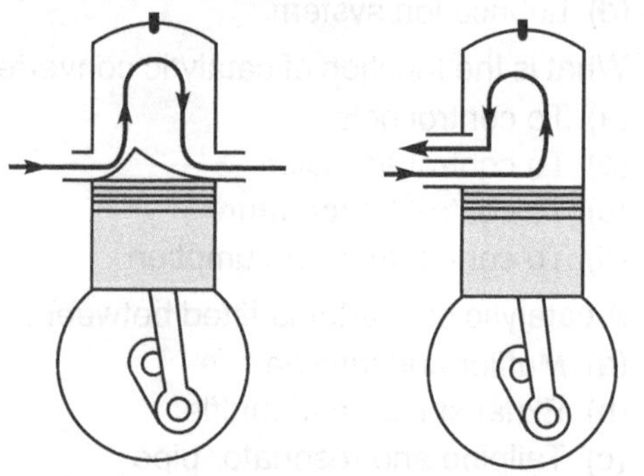

(a) Cross flow scavenge (b) Loop scavenge

Scavenging

24: The catalytic converter is the key part in the emission control system. It is designed to complete the combustion of exhaust gases as they exit the engine and before they reach the atmosphere. It is located between the engine and the tail pipe and its basic function is to convert the harmful components of the engine exhaust, such as carbon monoxide, to less harmful ones, such as carbon dioxide.

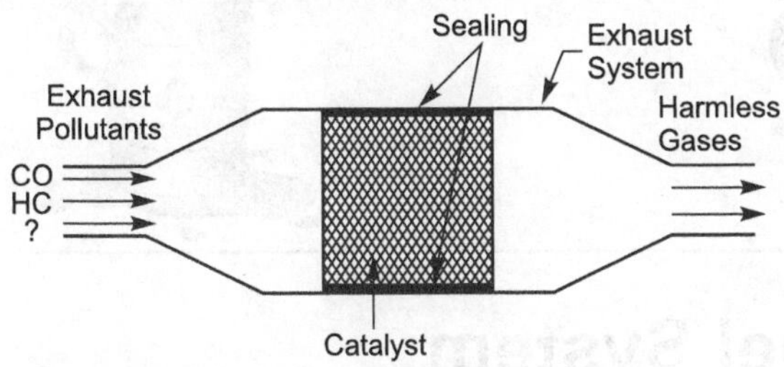

Catalytic converter

38: Selective catalytic reduction is a means of converting nitrogen oxides, also referred to as NO_x? with the aid of a catalyst into diatomic nitrogen, and water. A gaseous reductant, typically anhydrous ammonia, aqueous ammonia or urea, is added to a stream of flue or exhaust gas and is adsorbed onto a catalyst.

Module 12

Fuel System

MULTIPLE CHOICE QUESTIONS

1. Which part builds up fuel injection pressure in fuel injector?
   (a) Solenoid (b) Regulator
   (c) Common rail (d) Fuel pump
2. How is the pressure control valve operated in CRDI injectors?
   (a) Mechanically (b) Electronically
   (c) Manually (d) Hydraulically
3. Which electronic unit gives signal to operate IAC valve?
   (a) ECM (b) EDU
   (c) Solenoid (d) SCV
4. Where is the pressure discharge valve fitted in CRDI fuel system?
   (a) Common rail (b) Fuel pump
   (c) Injectors (d) Fuel filter
5. How many fuel chambers are there in HEUI?
   (a) One (b) Two
   (c) Three (d) Four
6. Which one of the following parts of diesel engine fuel tank is responsible for minimizing the slushing of fuel due to movement inside the tank?
   (a) Filler cap (b) Filler neck
   (c) Drain plug (d) Baffles
7. Which one of the following drives the fuel feed pump?
   (a) Engine flywheel (b) F.I.P. camshaft
   (c) Timing gears (d) Engine crankshaft
8. In a diesel engine, the feed pump is located on:
   (a) F. I. P. (b) Fuel tank
   (c) Fuel filter (d) Cylinder head
9. The hand priming device is put to use when the engine is:
   (a) Idling
   (b) Running at full speed
   (c) Not working
   (d) Running under full load
10. What is the name of part marked as 'X' in injector tester shown in the figure given below?

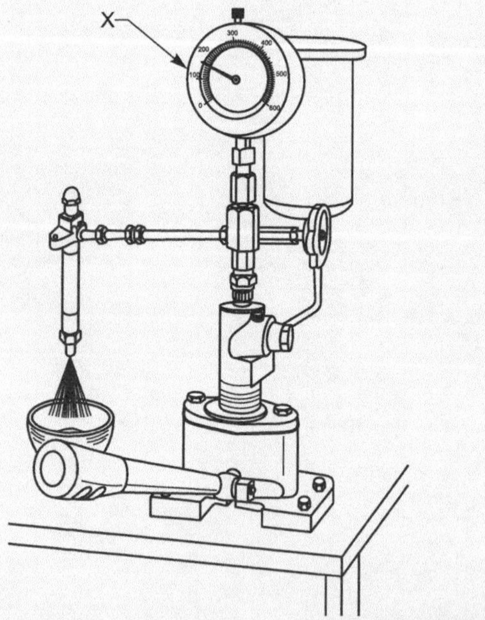

(a) Pressure gauge
(b) Vacuum gauge
(c) Temperature gauge
(d) Compression gauge

11. How much pressure is developed by the CRDI – diesel engine pump?
(a) 100 – 200 bar (b) 200 – 600 bar
(c) 600 – 1000 bar (d) 1000 – 1600 bar

12. Where is fuel injector fitted in an engine?
(a) Cylinder head (b) Cylinder block
(c) Head cover (d) Crankcase

13. What is the name of the dismantled component shown in the figure given below?

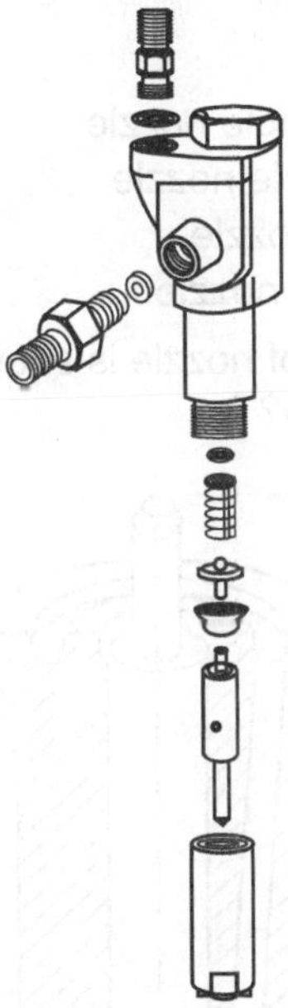

(a) Atomiser (b) FIP
(c) Heater plug (d) Feed pump

14. Which electronic device controls the engine system?
(a) Regulator (b) ECM
(c) Fuse (d) Switch

15. Which develops diesel pressure in CRDI system engine?
(a) ECM
(b) Injector
(c) Fuel tank
(d) High pressure pump

16. Which type of pump ensures in-built and uniform delivery to all injectors in diesel engine?
(a) Jerk type pump
(b) In-line pump
(c) Rotary type pump
(d) Servo type pump

17. What is the name of the governor in which the control rack is connected with counterweight floating lever?
(a) Mechanical governor
(b) Pneumatic governor
(c) Hydraulic governor
(d) Servo governor

18. Which fuel is related to Cetane Number?
(a) Petrol (b) Diesel
(c) Coal (d) Kerosene

19. Where from the diesel fuel is obtained?
(a) From crude oil
(b) From vegetable oil
(c) From animal oil
(d) From synthetic oil

20. Which fuel pipe line is provided between fuel filter and injectors?
(a) Suction pipe (b) Pressure pipe
(c) Vacuum pipe (d) Overflow pipe

21. During normal running, the necessary heat to ignite fuel in a diesel engine is obtained by using:
(a) Heater plugs
(b) Manifold heater
(c) A high compression ratio
(d) An un-cooled cylinder head

22. In an C.I. engine the fuel is injected when the piston:
(a) Is approaching t.d.c. at the end of compression stroke
(b) Has just passed t.d.c. at the start of power stroke
(c) Is approaching b.d.c. at the end of induction stroke
(d) Has just passed b.d.c. at the start of compression stroke

23. In a diesel engine combustion takes place at:

(a) Constant pressure
(b) Constant volume
(c) Constant temperature
(d) Constant temperature and pressure

24. To which one of the following parts of diesel engine the injector delivers fuel?

(a) Fuel tank (b) Fuel filter
(c) Engine cylinder (d) Governor

25. In diesel engine, fuel and air mix together in the:

(a) Carburettor
(b) Injector
(c) Combustion chamber
(d) Inlet port

26. A shallow depth combustion chamber has a cavity:

(a) In the cylinder head
(b) On the side of the cylinder
(c) In the crown of the piston
(d) Outside the cylinder

27. What type of nozzle is shown in the figure given below?

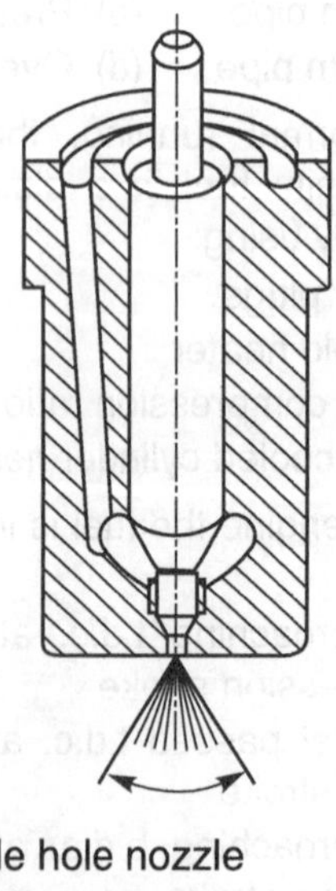

(a) Single hole nozzle
(b) Multi-hole nozzle
(c) Delay nozzle
(d) Pintle nozzle

28. What type of nozzle is shown in the figure given below?

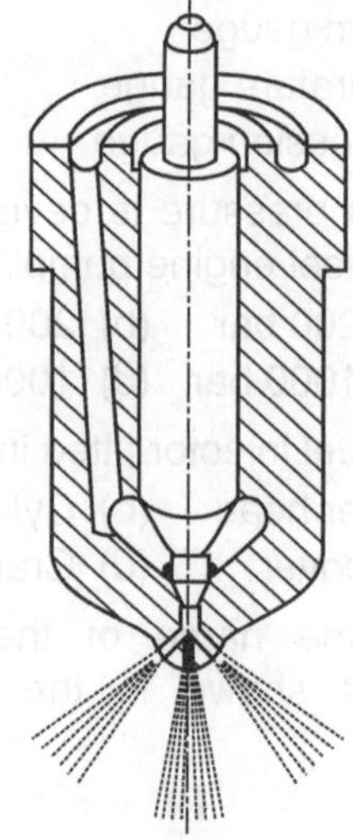

(a) Single hole nozzle
(b) Multi-hole nozzle
(c) Pintle nozzle
(d) Pintaux nozzle

29. What type of nozzle is shown in the figure given below?

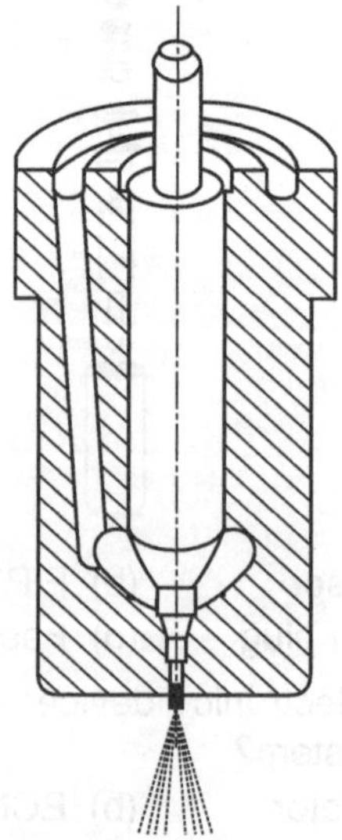

(a) Single hole nozzle
(b) Multi-hole nozzle
(c) Pintle nozzle
(d) Pintaux nozzle

30. Which part of an engine atomises the fuel into the cylinder?

(a) Injector

(b) FIP
(c) Governor
(d) Feed pump

31. Which type of fuel system is best suited for less fuel consumption, more power and reduction in exhaust emission?
(a) Inline pump system
(b) Rotary pump system
(c) Distributor pump system
(d) CRDI system

32. Which fuel system develops high diesel pressure by hydraulic energy?
(a) Inline FIP
(b) Distributor FIP
(c) CRDI fuel system
(d) HEUI fuel system

33. What is full form of 'HEUI'?
(a) Hydraulically Actuated Electronically Controlled Unit Injector
(b) Hydraulically Actuated Electrically Controlled Unit Injector
(c) Hydro-electric Controlled Unit Injector
(d) Hydraulic Effective Controlled Unit Injector

34. What is the name of the part marked 'X' in the figure given below?

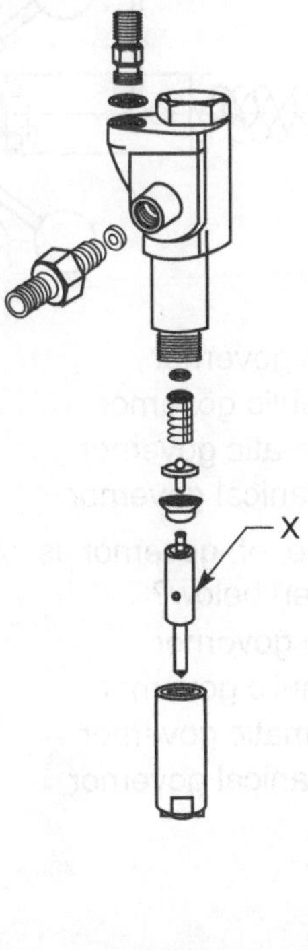

(a) Nozzle (b) Nozzle holder
(c) Spindle (d) Injector body

35. What is the name of the equipment shown in the figure given below?

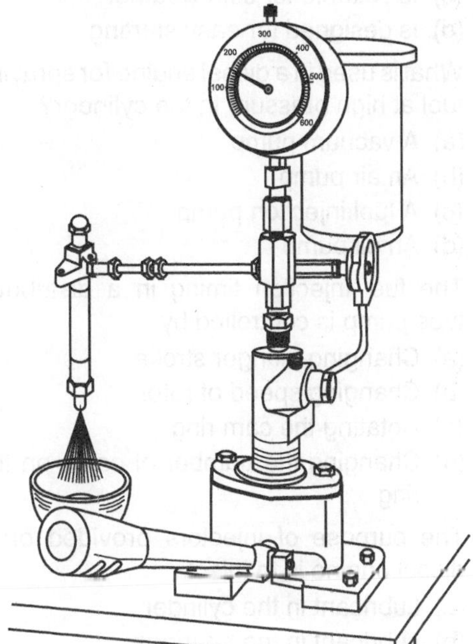

(a) FIP tester
(b) Injector tester
(c) Pump tester
(d) Compression tester

36. Which part of fuel injection pump compresses diesel?
(a) Governor (b) Rack rod
(c) Sleeve (d) Plunger

37. A glow plug is:
(a) A compression device
(b) A decompression device
(c) Fitted in the inlet manifold
(d) Screwed in the combustion chamber

38. Which one of the following in a diesel engine fuel system is located just before the injector?
(a) Fuel pump (b) Fuel tank
(c) Fuel filter (d) Lift pump

39. The special feature of pintle type nozzle is that it:
(a) Has a single hole
(b) Has an extended stem to form a pin
(c) Is suitable for cold weather
(d) Is designed for easy starting

40. What is used in a diesel engine for spraying fuel at high pressure in the cylinder?
(a) A vacuum pump
(b) An air pump
(c) A fuel injection pump
(d) An oil pump

41. The fuel injection timing in a distributor type pump is controlled by:
(a) Changing plunger stroke
(b) Changing speed of rotor
(c) Rotating the cam ring
(d) Changing the number of cams on the ring

42. The purpose of injectors provided on a diesel engine is to inject:
(a) Lubricant in the cylinder
(b) Lubricant in the bearings
(c) Fuel in the cylinders
(d) Coolant in the cylinder jackets

43. A governor is a:
(a) Temperature-sensitive device
(b) Pressure-sensitive device
(c) Speed-sensitive device
(d) Vacuum-sensitive device

44. Which one of the following works in conjunction with the governor?
(a) Coolant pump
(b) Fuel injection pump
(c) Lubricant pump
(d) Fuel feed pump

45. A governor, which provides perfect speed regulation with zero speed drop is called:
(a) Isobar governor
(b) Isotherm governor
(c) Isocline governor
(d) Isochronous governor

46. Which one of the following types of governors is also called spring-loaded centrifugal governor?
(a) Mechanical governor
(b) Hydraulic governor
(c) Pneumatic governor
(d) Servo governor

47. Venturi unit and diaphragm unit are the two main parts of which governor?
(a) Mechanical governor
(b) Hydraulic governor
(c) Pneumatic governor
(d) Servo governor

48. The operation of which governor is effective only if the system is air-tight?
(a) Mechanical governor
(b) Hydraulic governor
(c) Pneumatic governor
(d) Servo governor

49. What type of governor is shown in the figure given below?

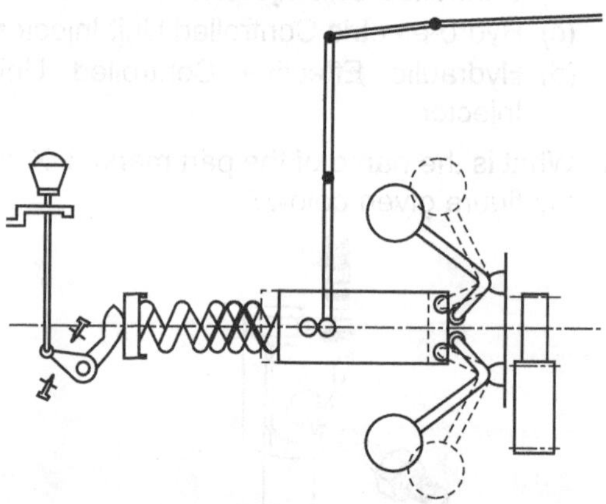

(a) Servo governor
(b) Hydraulic governor
(c) Pneumatic governor
(d) Mechanical governor

50. What type of governor is shown in the figure given below?
(a) Servo governor
(b) Hydraulic governor
(c) Pneumatic governor
(d) Mechanical governor

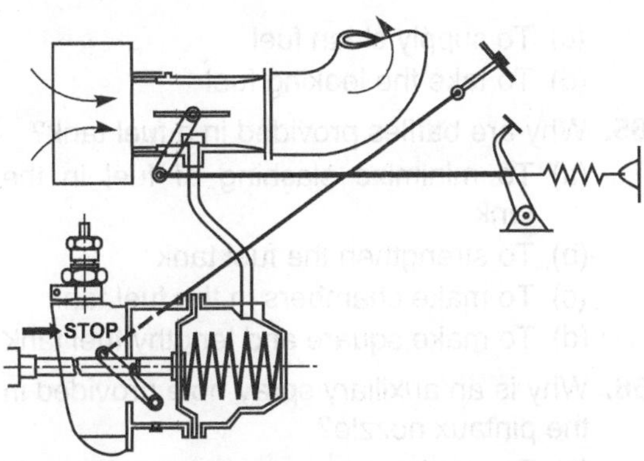

51. What is the name of the part marked 'X' in the figure given below?

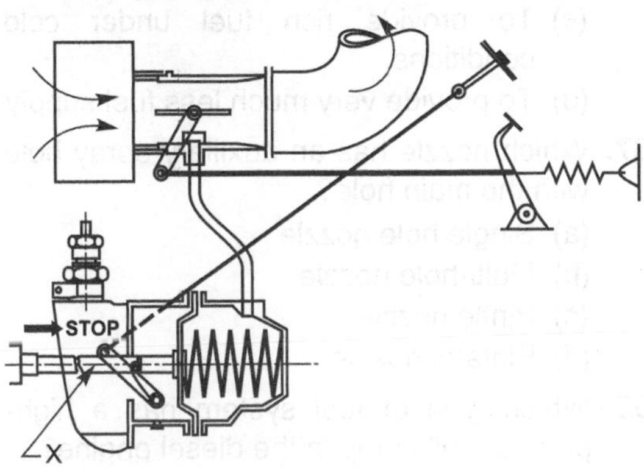

(a) Control rack
(b) Stop lever
(c) Diaphragm
(d) Vacuum chamber

52. In a pneumatic governor, what separates the vacuum chamber from the atmosphere?
(a) Control rack
(b) Stop lever
(c) Diaphragm
(d) Vacuum chamber

53. Which part is connected in-between the inlet manifold and low-pressure chamber of pneumatic governor?
(a) Accelerator cable
(b) Vacuum tube
(c) Stop lever cable
(d) Throttle lever

54. How is the quantity of fuel delivered is varied in a running diesel engine?
(a) By plunger (b) By control sleeve
(c) By control rack (d) By injection

55. Where the fuel feed pump of a diesel engine is usually mounted?
(a) Fuel filter
(b) Fuel gallery
(c) Fuel injector
(d) Fuel injection pump

56. What is the name of the part marked 'X' in the figure given below?

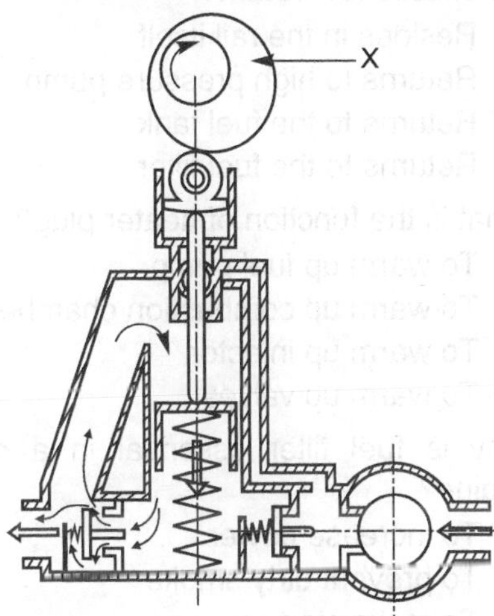

(a) Cam (b) Barrel
(c) Spindle (d) Plunger

57. What is the name of the part marked 'X' in the figure given below?

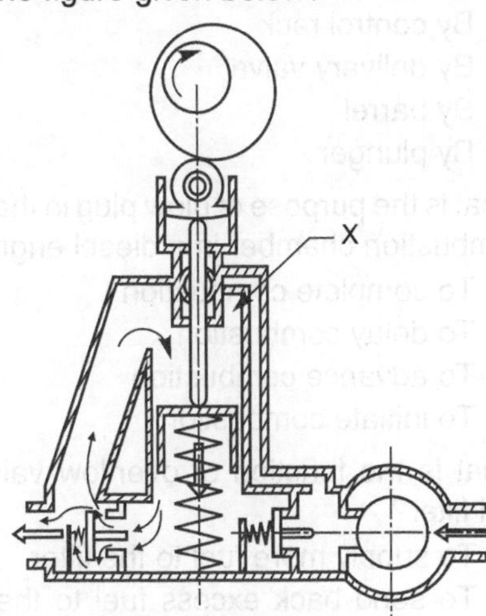

(a) Suction valve
(b) Delivery valve
(c) Suction chamber
(d) Pressure chamber

58. Which fuel feed system reduces un-burnt hydrocarbon by more than 50%?
(a) Inline fuel injection system
(b) Distributor fuel injection system
(c) Electronic control system
(d) Air blast fuel injection system

59. In CRDI engine fuel system, where does the excess fuel return?
(a) Resides in the rail itself
(b) Returns to high pressure pump
(c) Returns to the fuel tank
(d) Returns to the fuel filter

60. What is the function of heater plug?
(a) To warm up fuel pump
(b) To warm up combustion chamber
(c) To warm up injector
(d) To warm up valves

61. Why is fuel filter essential in a diesel engine?
(a) To increase power
(b) To prevent dirty smoke
(c) To make starting easy
(d) To polish nozzle and FIP to mirror finish

62. How is the pressure maintained in the high pressure pipeline of FIP?
(a) By control rack
(b) By delivery valve
(c) By barrel
(d) By plunger

63. What is the purpose of glow plug in the pre-combustion chamber in a diesel engine?
(a) To complete combustion
(b) To delay combustion
(c) To advance combustion
(d) To initiate combustion

64. What is the function of overflow valve in fuel filter?
(a) To supply more fuel to the filter
(b) To send back excess fuel to the fuel tank
(c) To supply clean fuel
(d) To take the leaking fuel

65. Why are baffles provided in a fuel tank?
(a) To minimize slashing of fuel in the tank
(b) To strengthen the fuel tank
(c) To make chambers in the fuel tank
(d) To make square and lengthy fuel tank

66. Why is an auxiliary spray hole provided in the pintaux nozzle?
(a) To assist easy starting under cold conditions
(b) To assist stop under cold conditions
(c) To provide rich fuel under cold conditions
(d) To provide very much less fuel supply

67. Which nozzle has an auxiliary spray hole with the main hole?
(a) Single hole nozzle
(b) Multi-hole nozzle
(c) Pintle nozzle
(d) Pintaux nozzle

68. Which type of fuel system has a high-pressure oil pump in the diesel engine?
(a) Inline FIP
(b) Rotary FIP
(c) CRDI fuel system
(d) HEUI fuel system

69. What controls the minimum and maximum speed of the diesel engine?
(a) FIP
(b) Governor
(c) Injector
(d) Feed pump

70. How much time does it take to give signal to ECM after ignition is switched on?
(a) One second (b) Two second
(c) Three second (d) Four second

71. What is the effect of detonation in a diesel engine?
(a) Knocking sound in the engine
(b) More engine power
(c) High engine speed
(d) Sudden stoppage of the engine

72. In order to eliminate knocking in a compression ignition engine, there should be:
(a) Short delay period
(b) Late auto-ignition
(c) Low compression ratio
(d) High self-ignition temperature of fuel

73. High sulphur content in diesel oil, if used in diesel engine will lead to:
(a) Production of highly corrosive gases corroding cylinder walls and exhaust system
(b) Excessive engine wear
(c) Damaging both the storage tank and the engine
(d) Reduction in thermal efficiency

74. The output of a diesel engine can be increased by:
(a) Feeding more fuel
(b) Heating incoming air
(c) Scavenging
(d) Supercharging

75. Airlock in fuel line will cause:
(a) Late starting and erratic running of the engine
(b) Excessive delivery from FI pump
(c) Too high injector pressure
(d) High engine idling

76. What for is the electrical solenoid provided on some fuel injection pumps?
(a) To heat the incoming fuel
(b) To provide excess fuel for starting
(c) To stop the engine through ignition switch and cut off fuel
(d) For electrical governing

77. Injectors and heater plugs are located in:
(a) Cylinder block (b) Cylinder head
(c) Crankcase (d) Valve

78. In a fuel tank, why vent hole is provided either in filler neck or in the cap?
(a) To maintain vacuum in the tank above the fuel
(b) To maintain atmospheric pressure in the tank above the fuel
(c) To maintain fuel vapour in the tank above the fuel
(d) To maintain fuel evaporation in the tank above the fuel

79. Which type of FIP has a single pumping element, which supplies fuel to all cylinders?
(a) Distributor pump
(b) Inline pump
(c) Feed pump
(d) CRDI pump

80. The process of testing a FIP for the accuracy of its supplying fuel at correct intervals is termed as ....
(a) Calibrating (b) Phasing
(c) Pumping (d) Servicing

81. Which type of engine has common rail direct fuel injection system?
(a) Steam engine (b) Diesel engine
(c) Turbine engine (d) Wankel engine

82. How is the CRDI injectors' pressure control valve operated?
(a) Mechanically (b) Electronically
(c) Manually (d) Hydraulically

83. Where is the pressure discharge valve fitted in CRDI fuel system?
(a) Common rail (b) Fuel pump
(c) Injectors (d) Fuel filter

84. Which develops diesel pressure in CRDI fuel system?
(a) ECM
(b) Injector
(c) Fuel tank
(d) High pressure pump

85. What process is involved in the fuel bleeding system?
(a) Burning air (b) Pumping air
(c) Removing air (d) Pressurising air

86. Where is the delivery pipe of fuel feed pump connected to?
(a) To the fuel tank
(b) To the fuel filter
(c) To the fuel gallery
(d) To the fuel injection pump

87. When do we use the hand priming pump?
(a) When the engine is at rest
(b) When the engine is at full load
(c) When the engine is at full speed
(d) When the engine is at idling speed

88. What is the effect of airlock in the fuel system?
(a) Hard running of the engine
(b) Erratic running of the engine
(c) Smooth running of the engine
(d) Continuous running of the engine

ANSWERS

1. (a)	2. (b)	3. (a)	4. (a)	5. (b)	6. (d)	7. (b)	8. (a)	9. (c)	10. (a)
11. (d)	12. (a)	13. (a)	14. (b)	15. (d)	16. (c)	17. (a)	18. (b)	19. (a)	20. (d)
21. (c)	22. (a)	23. (a)	24. (c)	25. (c)	26. (c)	27. (d)	28. (b)	29. (a)	30. (a)
31. (d)	32. (d)	33. (a)	34. (a)	35. (b)	36. (d)	37. (d)	38. (c)	39. (b)	40. (c)
41. (c)	42. (c)	43. (c)	44. (b)	45. (d)	46. (a)	47. (c)	48. (c)	49. (d)	50. (c)
51. (a)	52. (c)	53. (b)	54. (c)	55. (d)	56. (a)	57. (d)	58. (c)	59. (c)	60. (b)
61. (b)	62. (b)	63. (d)	64. (b)	65. (a)	66. (a)	67. (d)	68. (c)	69. (b)	70. (b)
71. (a)	72. (a)	73. (a)	74. (d)	75. (a)	76. (c)	77. (b)	78. (b)	79. (a)	80. (b)
81. (b)	82. (b)	83. (a)	84. (d)	85. (c)	86. (b)	87. (a)	88. (b)		

EXPLANATORY NOTES

1: The fuel injectors are each fitted with one solenoid valve.

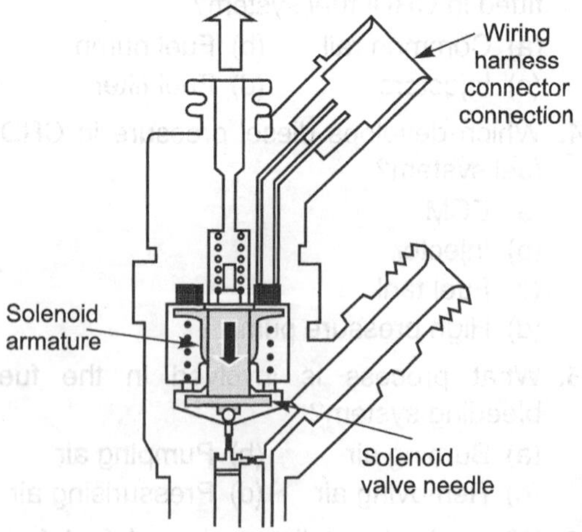

Injector

2: **CRDI:** Common Rail Direct Injection

3: **What does the Air Control Valve Do?** The IAC mainly controls the amount of air that is taken into the engine for combustion while the engine is running. The car's computer system monitors inputs from a variety of sensors which provides information on engine speed, airflow, fuel combustion, and much more. The engine computer will actuate a control motor on the IAC that will allow more or less air into the engine which in turn will control the engine speed.

ECM (Electronic Control Module): It analyzes or calculates the information which has come from the sensors and gives the instruction to the actuators.

EDC (Electronic Diesel Control): SCV: Suction Control Valve

5: HEUI (Hydraulically Actuated Electronically Controlled Unit Injector) is divided in two sections. One is low pressure fuel chamber. Another one is high pressure oil chamber.

12: Each fuel injector is fitted in the cylinder head above each combustion chamber.

16: The inline pump has a plunger and barrel assembly for each cylinder of the engine. The rotary type of fuel injection pump has a single pumping element, which supplies fuel to all the cylinders.

26: There are two different types of combustion chambers for C. I. Engines, i.e. direct injection type and indirect injection type. There are four different designs in direct injection type combustion chambers, i.e. shallow depth chamber, hemispherical chamber, cylindrical chamber and toroidal chamber.

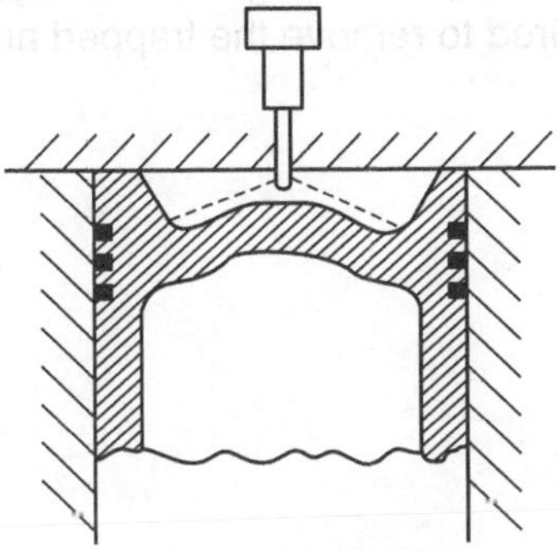

Shallow depth combustion chamber

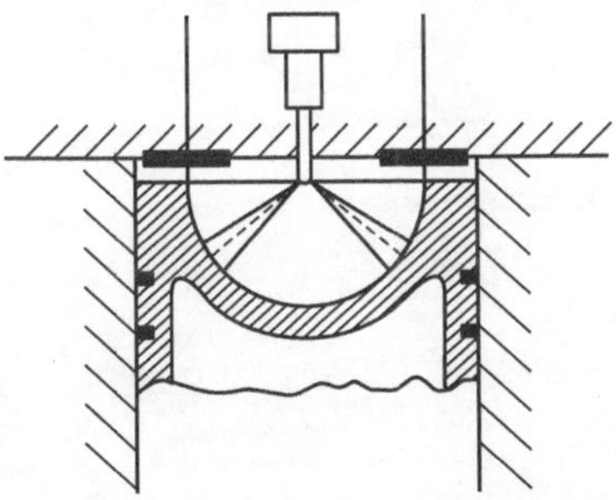

Hemispherical combustion chamber

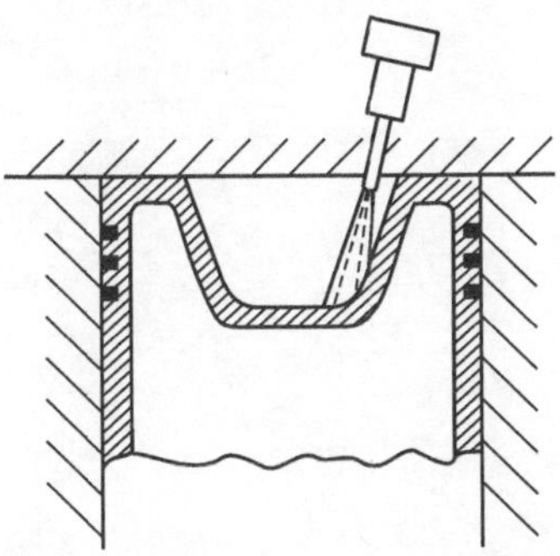

Cylindrical combustion chamber

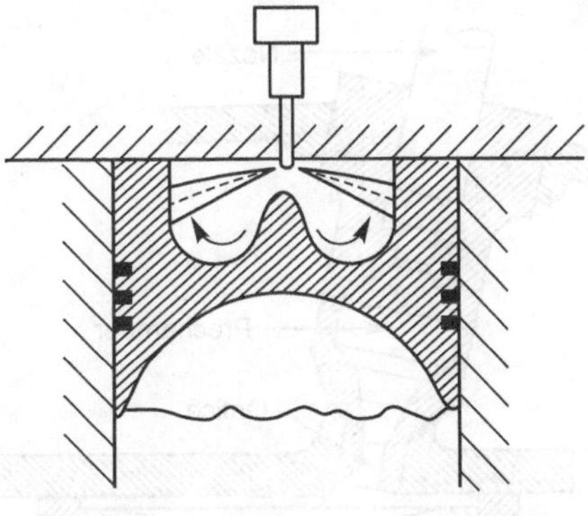

Toroidal combustion chamber

There are three designs in indirect type combustion chambers, i.e. swirl chamber, pre-combustion chamber and air cell chamber.

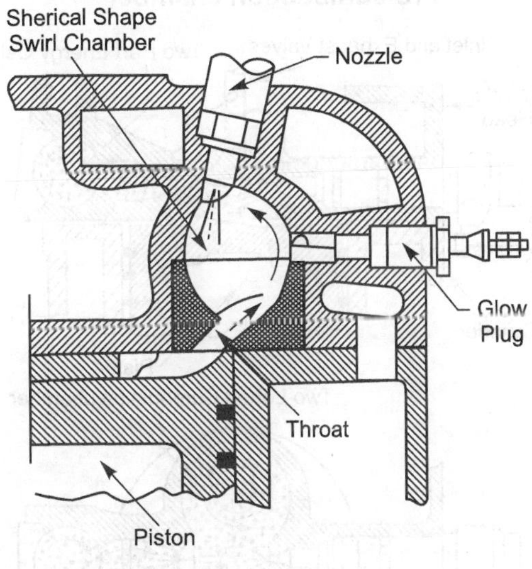

Swirl combustion chamber

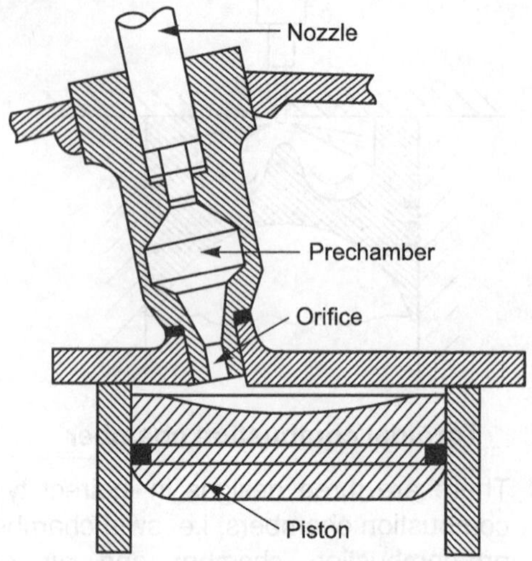

Pre-combustion chamber

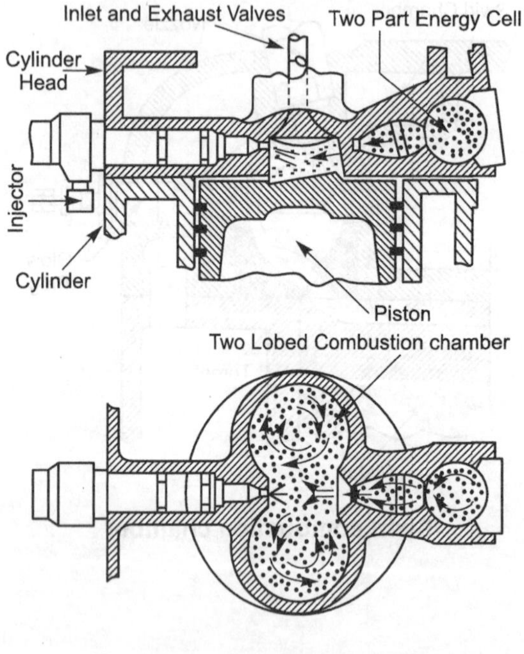

Air cell combustion chamber

80: The fuel system is designed to introduce fuel into the engine cylinders at the correct time, at correct high pressure, the correct quantity and correctly atomised. Phasing is the setting of the pump according to injection timing.

85: When you change a diesel fuel filter, run out of fuel or disturb the fuel system, air is trapped. When you try to start the engine this air acts as a lock, preventing the normal supply of fuel into the cylinder. Therefore, bleeding of the system is required to remove the trapped air.

Module 13

Marine and Stationary Engine

MULTIPLE CHOICE QUESTIONS

1. Air starting system is used for the engine in:
   (a) Cars (b) Bikes
   (c) Trucks (d) Marine
2. Which system includes an accumulator?
   (a) Cooling system
   (b) Starting system
   (c) Electrical system
   (d) Hydraulic system
3. Which is the commonly used cooling medium in marine engines?
   (a) Fan (b) Coolant
   (c) Seawater (d) Cooling water
4. Which fuel system is adopted in modern marine engines?
   (a) CRDI
   (b) In-line pump
   (c) Rotary pump
   (d) Reciprocating pump
5. What type of drive is shown in the figure given below?

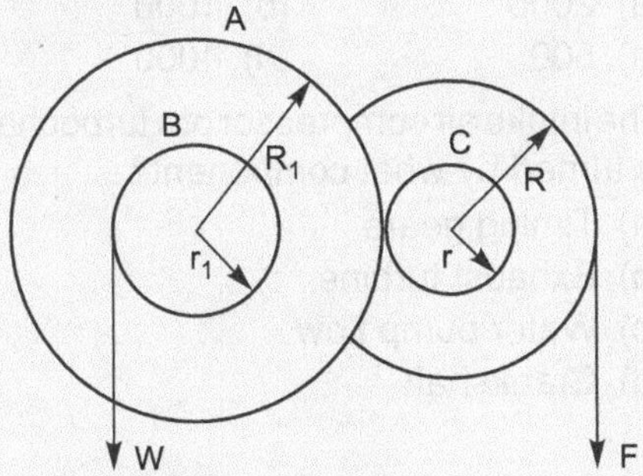

   (a) Electromagnetic
   (b) Single reduction
   (c) Double reduction
   (d) Hydraulic fluid coupling
6. Name the transmission device shown in the figure given below.

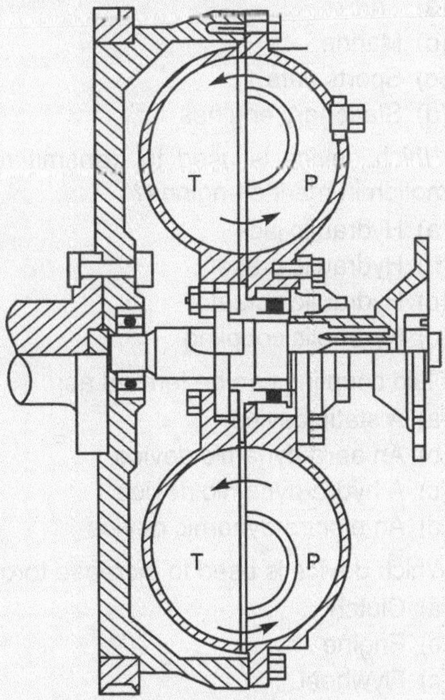

   (a) Fluid coupling
   (b) Electromagnetic coupling
   (c) Universal coupling
   (d) Flange coupling

7. Name the engine shown in the figure given below.

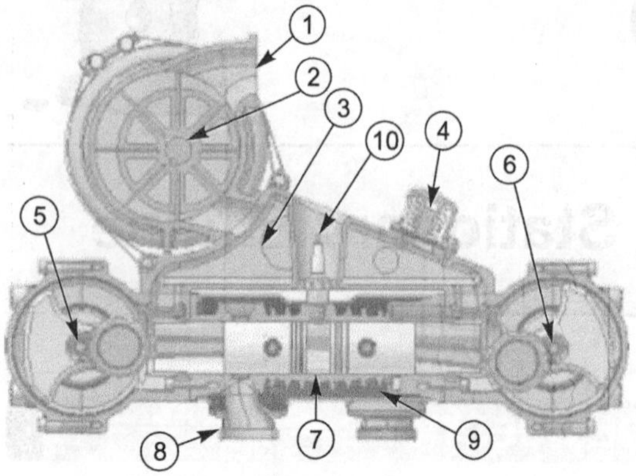

(a) Double-acting engine
(b) Opposed piston engine
(c) Single acting reciprocating engine
(d) Double acting reciprocating engine

8. The auxiliary engine starting system is used in:
(a) Trucks
(b) Marine
(c) Sports cars
(d) Stationary engines

9. Which device is used to transmit rotary motion in marine engines?
(a) Hydraulic jack
(b) Hydraulic brake
(c) Hydraulic actuator
(d) Hydraulic coupling

10. Fluid coupling can be termed as:
(a) A static device
(b) An aero-dynamic device
(c) A hydro-dynamic device
(d) An electro-dynamic device

11. Which device is used to increase torque?
(a) Clutch
(b) Engine
(c) Flywheel
(d) Gear reduction drive

12. Which one of the following fuel injection systems was originally used in large stationary marine engines?
(a) Air injection
(b) Airless injection
(c) Mechanical injection
(d) Solid injection

13. Diesel engine heat exchanger tubes are made of good conductor of heat material. What is the material used for heat exchanger tubes?
(a) Aluminium (b) Steel
(c) Brass (d) Stainless steel

14. Marine engines are:
(a) Air cooled
(b) Water cooled
(c) Oil cooled
(d) Forced convection cooled

15. The starter motor of an engine is meant to .... the engine.
(a) Crank (b) Start
(c) Stall (d) Idle

16. Each movement of the piston in the cylinder is called:
(a) Compression (b) Exhaust
(c) Stroke (d) Shift

17. Engine revolutions per minute (RPM) measure the rotations of what component?
(a) Camshaft (b) Crankshaft
(c) Starter (d) Connecting rod

18. What is the minimum number of cylinders required for a four-stroke engine to ensure no interruptions in output power? (i.e. at least one cylinder in power stroke).
(a) Five (b) Three
(c) One (d) Four

19. A four-stroke, four-cylinder engine is running at 1000 RPM. How many power strokes (ignitions) occur during one minute?
(a) 2000 (b) 1000
(c) 500 (d) 4000

20. The intake air compressor of a turbocharger is turned by what component?
(a) Timing gears
(b) Exhaust turbine
(c) Water pump flow
(d) Crankshaft

21. If the temperature of intake air in I.C. engine is lowered, then its efficiency will:
(a) Increase
(b) Decrease
(c) Remain the same
(d) Increase up to a certain limit and then decrease

22. Heat engine uses .... energy to produce mechanical work.
(a) Thermal (b) Fluid
(c) Kinetic (d) Electric

23. The energy of expanding gas in the engine cylinder is transferred by the piston to the connecting rod through:
(a) Crankpin (b) Gudgeon pin
(c) Bearing (d) Crankcase

24. Which of the following is not used in four-stroke compression-ignition (CI) engines?
(a) Fuel pump
(b) Spark plug
(c) Fuel injector
(d) Inlet and outlet valves

25. In two-stroke engine, there is one power stroke in .... of crankshaft rotation.
(a) 90° (b) 180°
(c) 270° (d) 360°

26. In four-stroke engine which of the following is also known as power stroke?
(a) Suction stroke
(b) Compression stroke
(c) Expansion stroke
(d) Exhaust stroke

ANSWERS

1. (d)	**2.** (b)	**3.** (c)	**4.** (a)	**5.** (c)	**6.** (a)	**7.** (b)	**8.** (b)	**9.** (d)	**10.** (c)
11. (d)	**12.** (a)	**13.** (c)	**14.** (b)	**15.** (a)	**16.** (c)	**17.** (b)	**18.** (d)	**19.** (a)	**20.** (b)
21. (a)	**22.** (a)	**23.** (b)	**24.** (b)	**25.** (d)	**26.** (c)				

EXPLANATORY NOTES

1: A starter is what gets an engine going initially, before it continues under its own momentum. In cars and motorbikes, this is generally done with an electric starter. Large Marine Diesel Engines are started using high pressure compressed air.

To start an engine with an air starter, compressed air is forced into a cylinder, forcing the piston downwards. As the piston is forced down, the air is able to flow into the other cylinders, pushing those pistons downwards as well, continuing the process of starting the engine. As the pistons move, fuel is drawn into the cylinders and the engine begins running, just as it would if started with an electric system.

3: The figure given below shows the system used in cooling of marine engines. The system used is closed type with a water-to-water heat exchanger. The primary water (soft water) from the engines goes to the heat exchanger where heat is absorbed by secondary water (raw water). A bypass control is placed around the heat exchanger to control water temperature. After having been cooled in the heat exchanger, the primary water flows to the engine water jackets again. An expansion tank is used in this system to overcome the variation in water volume due to expansion and contraction by heating and cooling. The secondary water (raw water) taken from sea or river passes through the lubricating oil cooler and heat exchanger. It is then discharged back to the sea or river.

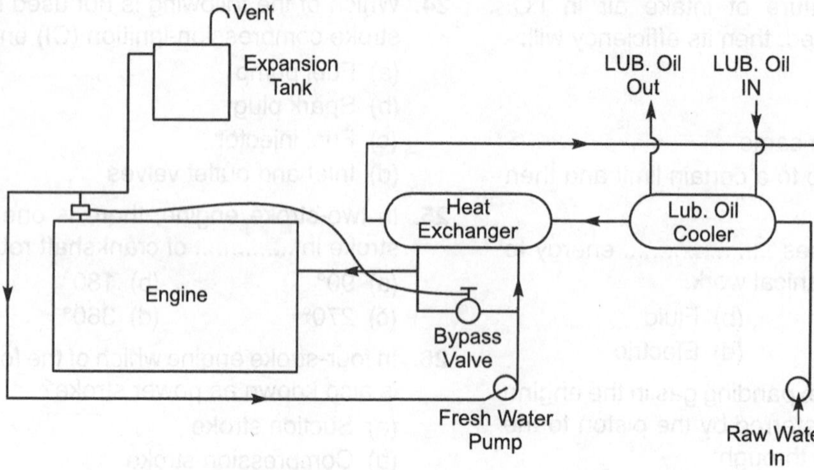

Water cooling system used in marine engines

10: Hydrodynamics is a branch of physics that deals with the motion of fluids and the forces acting on solid bodies immersed in fluids and in motion relative to them.

Module 14

Emission Control System

MULTIPLE CHOICE QUESTIONS

1. Where from an engine, hydrocarbon emission is released directly to the atmosphere?
(a) From crankcase and exhaust system
(b) From fuel tank and carburettor
(c) From fuel tank and crankcase
(d) From fuel tank and exhaust system

2. Which among the following engines emits more amount of nitrogen oxides (NOx)?
(a) Spark ignition engine
(b) Compression ignition engine
(c) Two-stroke engine
(d) LPG engine

3. Which is a greenhouse effect gas?
(a) O_2 (b) CO
(c) CO_2 (d) N_2

4. Which pollutant is released more from diesel engine during weak compression?
(a) CO (b) Particulate matter
(c) HC (d) NOx

5. Which is a non-polluting gas?
(a) CO (b) NO
(c) HC (d) O_2

6. Which is a colourless, odourless, tasteless and highly toxic gas?
(a) CO (b) CO_2
(c) NOx (d) SOx

7. What is the full form of acronym 'DEF' in the contest of selective catalytic reduction?
(a) Diesel Engine Fluid
(b) Diesel Exhaust Fluid
(c) Diesel Emission Fluid
(d) Diesel Engine Fuse

8. What is the full form of acronym 'SCR' in the context of emission control system?
(a) Silicon Controlled Rectifier
(b) Selective Catalytic Reduction
(c) Selective Controlled Rectifier
(d) Selective Controlled Reduction

9. What is the full form of acronym 'EGR' in the context of emission control system?
(a) Engine Gas Recirculation
(b) Exhaust Gas Recirculation
(c) Exhaust Gate Regulator
(d) Emission Gas Recirculation

10. Much of the heat in the exhaust gases may be recovered by the use of:
(a) Heat-exchanger (b) Muffler
(c) Manifold (d) Tail pipe

11. The deposit of carbon in the exhaust system will:
(a) Increase back pressure
(b) Reduce back pressure
(c) Have no effect on back pressure
(d) Result in fluctuating back pressure

12. Where the positive crankcase ventilation is fitted?
(a) Muffler and air cleaner
(b) Controller and air cleaner
(c) Feed pump and air cleaner
(d) Engine breather and air cleaner

13. What are catalytic converters used for?
(a) To control noise
(b) To control emission
(c) To control temperature
(d) To control fuel consumption

14. What is the function of EGR (Exhaust Gas Recirculation) valve?
(a) To reduce NOx (b) To reduce CO
(c) To reduce CO_2 (d) To reduce SO_2

15. What is the function of EVAP canister?
(a) To trap exhaust gas
(b) To trap fresh charge
(c) To trap leaking fuel
(d) To trap fuel vapour

16. What will happen if diesel is used in a petrol engine?
(a) Engine won't run
(b) Engine will run more efficiently
(c) Engine will run at higher speed
(d) Engine will explode

17. The EVAP canister outlet is connected to:
(a) Inlet manifold
(b) Exhaust manifold
(c) Catalytic converter
(d) Fuel tank

18. Which device changes carbon monoxide in the exhaust gas into carbon dioxide?
(a) Canister
(b) Muffler
(c) Catalytic converter
(d) Exhaust gas regulator

19. What is the effect of high sulphur content in fuel?
(a) Incomplete combustion
(b) Pitting in the engine parts
(c) Fungus on the cylinder walls
(d) Corrosive wear of engine parts

20. Which part of engine exhaust system is also called 'exhaust silencer'?
(a) Exhaust valve
(b) Muffler
(c) Tail pipe
(d) Exhaust manifold

21. Which type of engine uses SCR for emission control?
(a) Petrol engine (b) Diesel engine
(c) LPG engine (d) CNG engine

22. When does fresh charge escape from petrol engine exhaust?
(a) During valve lead
(b) During valve open
(c) During valve lag
(d) During valve overlap

23. Which type of fuel is recommended for engines having catalytic converter installed in it?
(a) Leaded petrol
(b) Unleaded petrol
(c) High speed diesel
(d) Low speed diesel

24. In which type of emission control system the blow by gases are fed into the inlet manifold of a running engine?
(a) Exhaust gas recirculation
(b) Crankcase ventilation
(c) Positive crankcase ventilation
(d) Catalytic conversion

25. In modern vehicles, what controls the EGR valve?
(a) Engine vacuum controlled valve
(b) Vacuum regulated valve
(c) Linear electronic controlled valve
(d) Solenoid controlled valve

26. What does 'EPA' refer to in respect of environment?
(a) Environmental Protection Agency
(b) Environmental Provision Act
(c) Environmental Protection Act
(d) Environmental Provisions Agency

27. What is the full form the acronym 'DPF' in the context of exhaust emission system?
(a) Diesel Pressure Filter
(b) Diesel Primary Filter
(c) Diesel Particulate Filter
(d) Direct Particulate Filter

28. Which type of engine uses EVAP canister?
(a) Diesel engine (b) Petrol engine
(c) LPG engine (d) CNG engine

29. What is used in EVAP canister to absorb fuel vapours?
(a) Banian cloth
(b) Cotton roll
(c) Paper filter
(d) Activated charcoal

30. What causes emission of particulate matter?
(a) Complete combustion
(b) Incomplete combustion
(c) Supply of excess air
(d) Atmospheric mixing

31. How can near zero emission in diesel engine be achieved?
(a) Combo of PCV and EGR
(b) Combo of EGR and SCR
(c) Combo of PCV and 3-way catalytic converter
(d) Combo of DPF and EGR

ANSWERS

1. (a)	**2.** (b)	**3.** (c)	**4.** (b)	**5.** (d)	**6.** (a)	**7.** (b)	**8.** (b)	**9.** (b)	**10.** (a)
11. (a)	**12.** (d)	**13.** (b)	**14.** (a)	**15.** (d)	**16.** (a)	**17.** (a)	**18.** (c)	**19.** (c)	**20.** (b)
21. (b)	**22.** (d)	**23.** (b)	**24.** (c)	**25.** (c)	**26.** (d)	**27.** (c)	**28.** (b)	**29.** (d)	**30.** (b)
31. (b)									

EXPLANATORY NOTES

General Remark: Modern diesel emission controls can reduce carbon monoxide, hydrocarbons, particulate matter, oxides of nitrogen, and toxic air pollutants. Widespread use of these control systems can significantly improve air quality.Controlling harmful emissions is especially important when diesel engines are operated in confined or restricted space where workers may be exposed to high exhaust concentrations.

7: Selective Catalytic Reduction (SCR) is an advanced active emissions control technology system that injects a liquid-reductant agent through a special catalyst into the exhaust stream of a diesel engine. The reductant source is usually automotive-grade urea, otherwise known as Diesel Exhaust Fluid (DEF).

9: The purpose of exhaust gas recirculation (EGR) system is to reduce NOx emissions that contribute to air pollution.

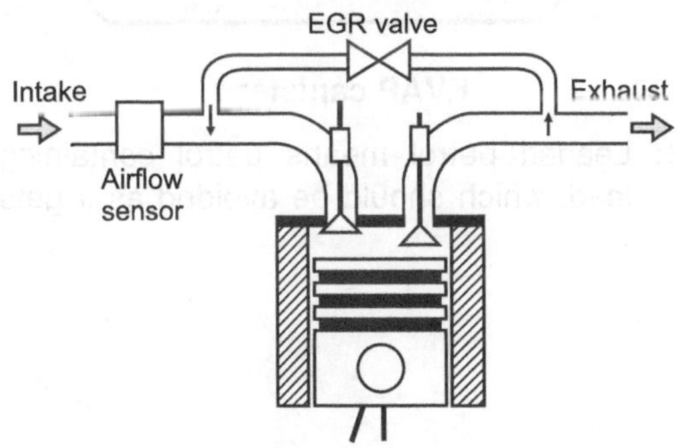

EGR valve

12: In an internal combustion engine, a crankcase ventilation system removes unwanted gases from the crankcase. The system usually consists of a tube, a one-way valve and a vacuum source. The unwanted gases, called "blow-by", are gases from the combustion chamber which have leaked past the piston rings.

PCV valve lifts the valve according to negative pressure at the inlet manifold to create appropriate ventilation for the crankcase.

15: The EVAP canister houses 1-2 pounds of activated charcoal that acts like a sponge for absorbing and storing fuel vapours until the purge valve opens and allows the vacuum of the engine intake to siphon the fuel vapours from the charcoal into the engine intake manifold.

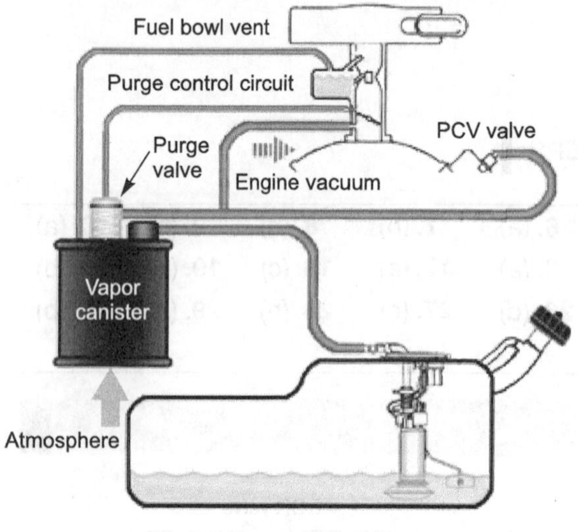

EVAP canister

23: Leaded petrol means petrol containing lead, which should be avoided as it gets emitted into the air and causes harmful effects on plants, humans and animals too, while unleaded petrol means petrol without lead, which is used now-a-days.

24: Positive Crankcase Ventilation (PCV) is a system that was developed to remove harmful vapours from the engine and to prevent those vapours from being expelled into the atmosphere.The PCV system does this by using manifold vacuum to draw vapours from the crankcase into the intake manifold.

28: The Evaporative Emission Control System (EVAP) is used to prevent gasoline vapours from escaping into the atmosphere from the fuel tank and fuel system.

29: Activated charcoal is a form of carbon processed to have small, low-volume pores that increase the surface area available for adsorption or chemical reactions.

31: PCV: Positive Crankcase Ventilation; EGR: Exhaust Gas Recirculation; SCR: Selective Catalytic Reduction; DPF: Diesel Particulate Filter. *3-way catalytic converters* are so named because they are capable of eliminating three pollutants – NOx, CO and HC.

Module 15

Charging and Starting System

MULTIPLE CHOICE QUESTIONS

1. Which one among the following is a strong electromagnetic switch in engine starting system?
   (a) Starter switch
   (b) Solenoid switch
   (c) Ignition switch
   (d) Starter push switch

2. What is the type of drive mechanism shown in the figure given below?

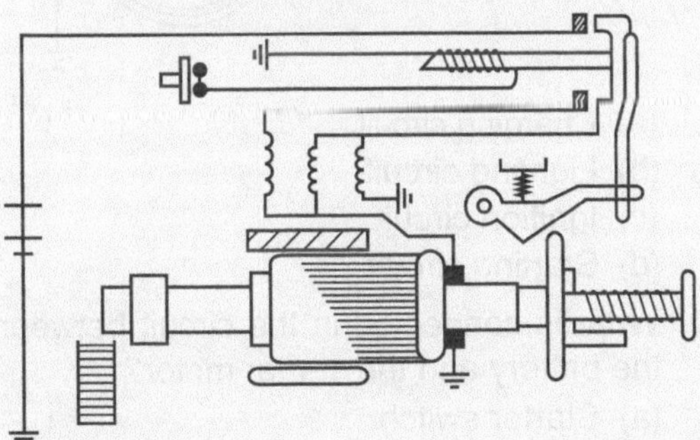

   (a) Overrunning clutch drive
   (b) Bendix drive
   (c) Axial drive
   (d) Non-axial type

3. What is the function of the regulator in an alternator?
   (a) To limit alternator field current as required
   (b) To permit current to flow in one direction only
   (c) To take away heat from the diodes
   (d) To increase current flow

4. Which part prevents back flow of current in an alternator?
   (a) Regulator
   (b) Rotor coil
   (c) Slip ring
   (d) Rectifier diode

5. What is connected to the starter motor shaft?
   (a) Drive pinion (b) Ring gear
   (c) Drive pulley (d) Drive coupling

6. The alternator is generally fitted at the engine's:
   (a) Front side (b) Rear side
   (c) Centre (d) Bottom side

7. In an alternator the magnetic field is produced in the:
   (a) Rotor (b) Stator
   (c) Frame (d) Regulator

8. Which type of drive mechanism slides the armature in the starter motor?
   (a) Bendix drive
   (b) Overrunning clutch drive
   (c) Axial drive
   (d) Non-axial drive

9. How many windings are provided in a solenoid switch?
   (a) 2 (b) 3
   (c) 4 (d) 5

10. What instrument is shown in the figure given below?

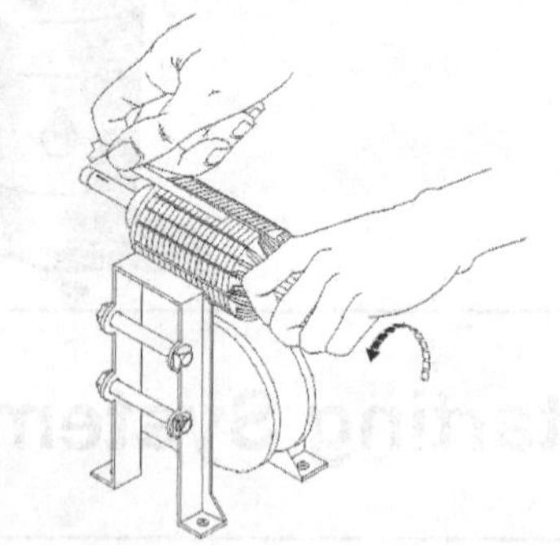

(a) Dial gauge (b) Growler
(c) Multi-meter (d) Hydrometer

11. Which instrument is used to check run-out of a commutator?
(a) Growler (b) Multi-meter
(c) Dial gauge (d) Vernier calliper

12. Which instrument is used to check wear of a commutator?
(a) Growler (b) Multi-meter
(c) Dial gauge (d) Vernier calliper

13. The magneto in an automobile is basically a:
(a) Transformer (b) D. C. Generator
(c) Capacitor (d) Magnetic circuit

14. The role of which part of a dynamo is to generate electromotive force?
(a) Commutator (b) Negative brush
(c) Positive brush (d) Armature

15. Minimum output of a dynamo is checked when:
(a) Battery is fully charged
(b) All the lights are ON and battery is fully discharged
(c) Electrical loads are with non-discharge
(d) Headlights ON with battery partially discharged

16. In a series coil in a shifter solenoid of starter motor builds magnetic field:
(a) During full voltage application
(b) When current flow is maximum at heavy cranking
(c) Only when the motor rotates at very low load
(d) When the motor runs at very high speed

17. What type of winding is best suitable to produce a high starting torque in starter motor?
(a) Series winding
(b) Shunt winding
(c) Compound winding
(d) Parallel winding

18. The starter motor of an engine is meant to .... the engine.
(a) Crank (b) Start
(c) Stall (d) Idle

19. What circuit is shown in the figure given below?

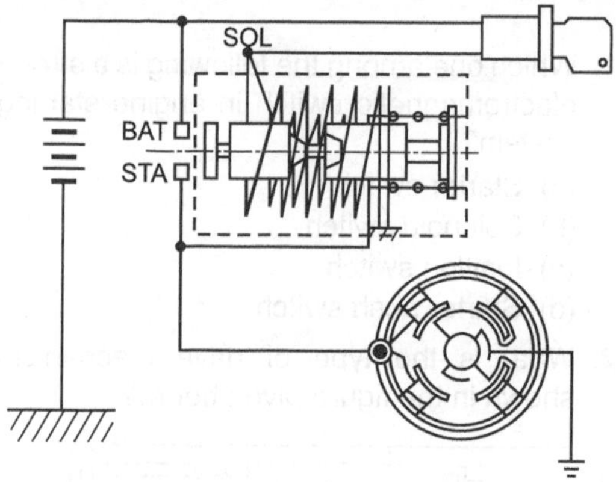

(a) Charging circuit
(b) Lighting circuit
(c) Ignition circuit
(d) Starting circuit

20. What is connected in the circuit between the battery and the starter motor?
(a) Starter switch
(b) Solenoid switch
(c) Plunger
(d) Ignition switch

21. What is the name of the part marked 'X' in the figure given below?
(a) Solenoid winding
(b) Battery
(c) Plunger
(d) Starter motor

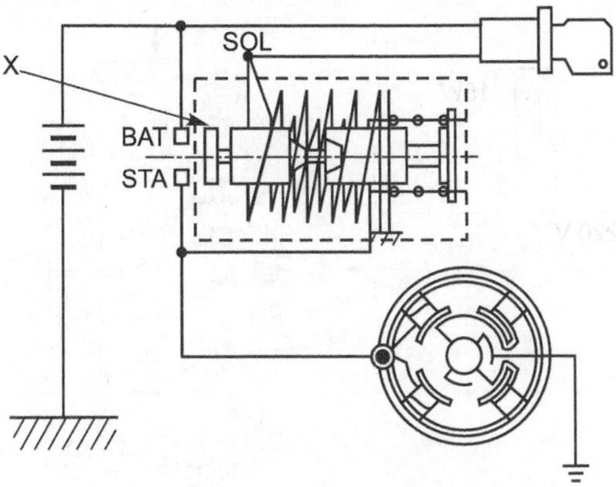

(a) Lighting circuit (b) Ignition circuit
(c) Starting circuit (d) Charging circuit

22. What is the name of the part marked 'X' in the figure given below.

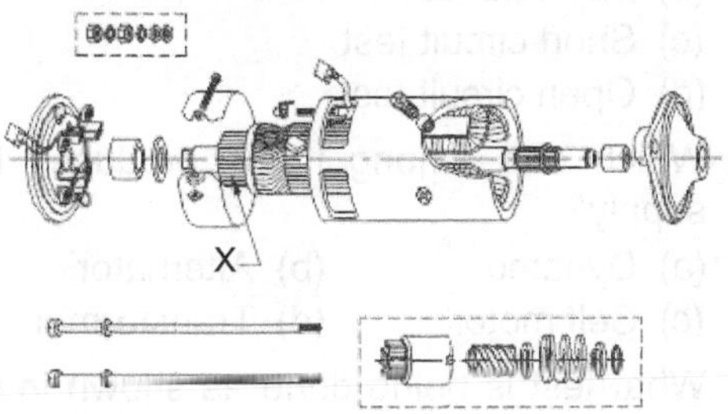

(a) Armature (b) Drive pinion
(c) Field coil (d) Commutator

23. In a starter motor, the carbon brushes are in contact with:
(a) Armature (b) Armature shaft
(c) End cover (d) Commutator

24. What circuit is shown in the figure given below?

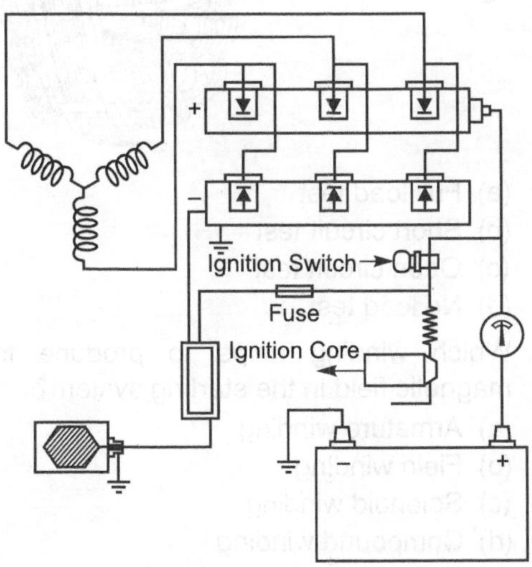

25. In an alternator, what is used to allow current to flow in one direction only?
(a) Field coil
(b) Armature
(c) Voltage regulator
(d) Diode

26. Which one among these produces AC supply?
(a) Dynamo (b) Alternator
(c) Starter motor (d) Transformer

27. Which part of an engine engages with the starter motor pinion?
(a) Timing gear
(b) Pump gear
(c) Flywheel ring gear
(d) Camshaft gear

28. What is the position of the pinion in the figure given below?

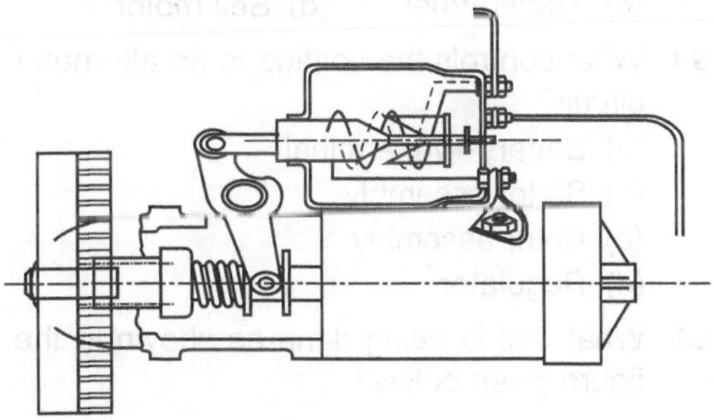

(a) Fully engaged
(b) Disengaged
(c) Partially engaged
(d) Fully disengaged

29. What is used to operate the pinion of the overrunning clutch drive?
(a) Starter switch
(b) Solenoid switch
(c) Ignition switch
(d) Starter push switch

30. What is the name of the instrument shown in the figure given below?
(a) Ammeter (b) Multimeter
(c) Hydrometer (d) Voltmeter

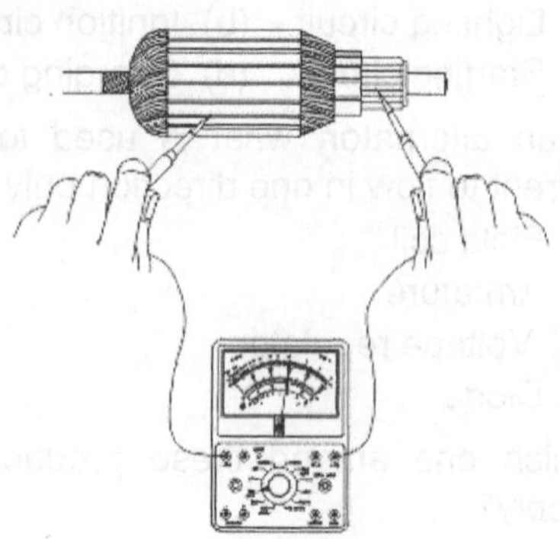

31. What device is used to maintain constant voltage in a charging circuit?
(a) Cut-out regulator
(b) Field regulator
(c) Voltage regulator
(d) Current regulator

32. Which device gives high output at low speed?
(a) Dynamo (b) Alternator
(c) Transformer (d) Self motor

33. What controls the voltage in an alternator circuit?
(a) Battery +ve terminal
(b) Stator assembly
(c) Rotor assembly
(d) Regulator

34. What test is being done as shown in the figure given below?

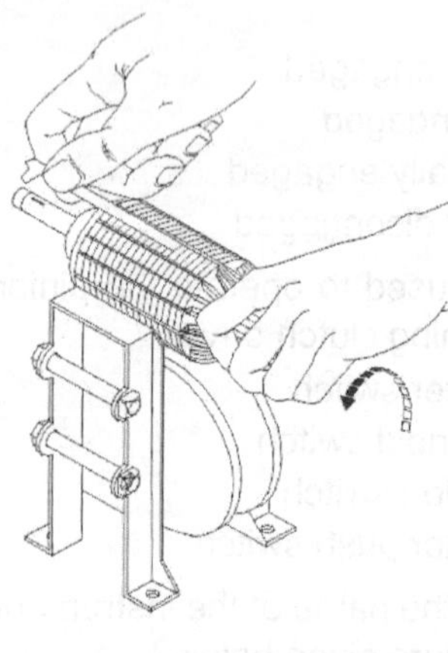

(a) Short circuit test (b) Open circuit test
(c) Continuity test (d) Insulation test

35. What test is being done as shown in the figure given below?

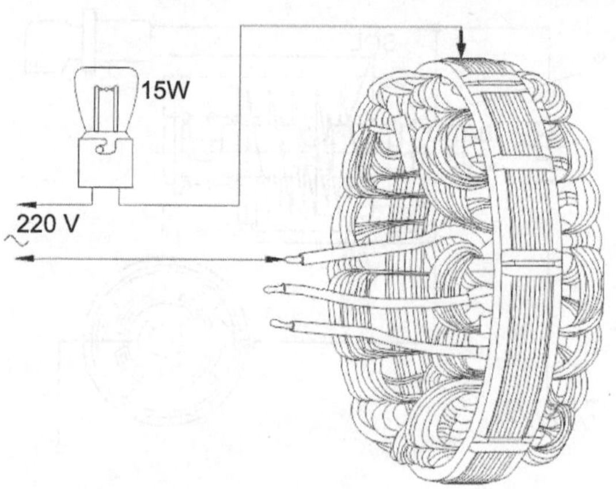

(a) No load test
(b) Full load test
(c) Short circuit test
(d) Open circuit test

36. Which one among these produces DC supply?
(a) Dynamo (b) Alternator
(c) Self motor (d) Transformer

37. What test is being done as shown in the figure given below?

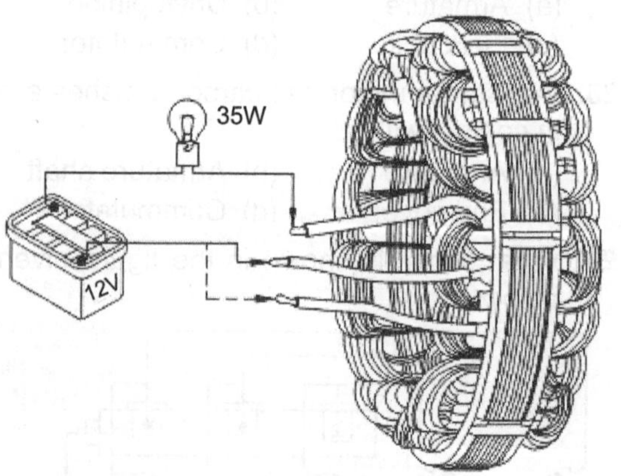

(a) Full load test
(b) Short circuit test
(c) Open circuit test
(d) No load test

38. Which winding helps to produce the magnetic field in the starting system?
(a) Armature winding
(b) Field winding
(c) Solenoid winding
(d) Compound winding

39. How does an alternator get the drive from the engine?
(a) By coupling (b) By gear
(c) By chain (d) By belt

40. Where are slip rings used?
(a) In dynamo (b) In alternator
(c) In self motor (d) In transformer

41. 41 Where is commutator used?
(a) In dynamo (b) In alternator
(c) In transformer (d) In ignition coil

42. What test is being done as shown in the figure given below?

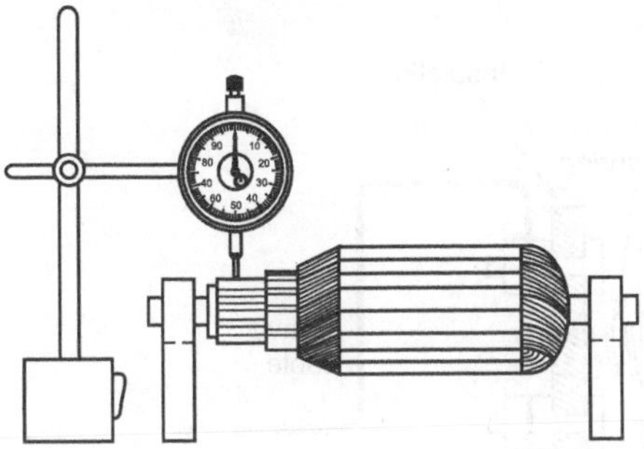

(a) Growler test
(b) Continuity test
(c) Insulation test
(d) Commutator run-out test

43. What test is being done as shown in the figure given below?

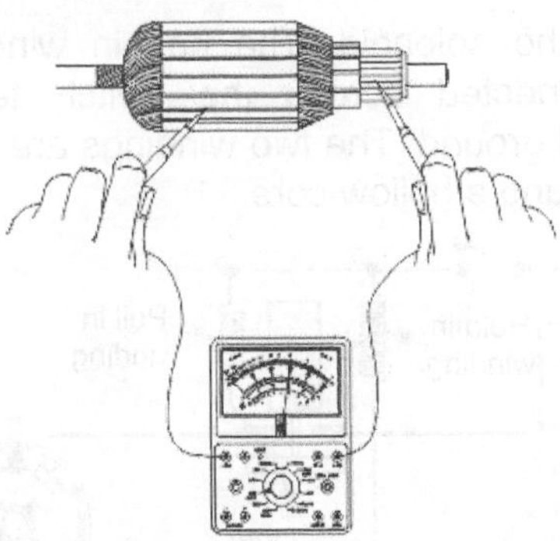

(a) Insulation test (b) Growler test
(c) Continuity test (d) Open circuit test

44. What is the probable cause for charge at high rate?
(a) Loose mounting
(b) Blown fuse wire
(c) Broken drive belt
(d) Voltage regulator winding open

45. What is the probable cause for a noisy alternator?
(a) Loose mounting
(b) Blown fuse wire
(c) Broken drive belt
(d) Voltage regulator winding open

46. What test is being done as shown in the figure given below?

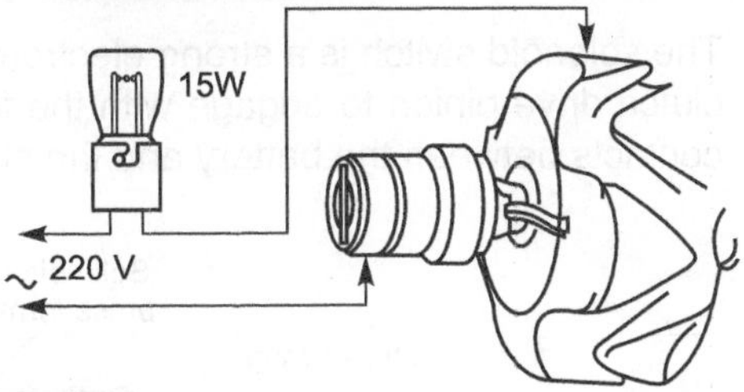

(a) Full load test (b) No load test
(c) Short circuit test (d) Open circuit test

47. A solenoid uses two coils. Their windings are called:
(a) Push-in and pull-out
(b) Pull-in and push-out
(c) Push-in and hold-out
(d) Pull-in and hold-in

48. If an engine cranks too slow to start, the problem may be caused by:
(a) Engine problem
(b) A faulty neutral start switch
(c) An open relay in the control circuit
(d) A damaged pinion gear

49. If the starter spins but does not crank the engine, the likely cause would be:
(a) A bad overrunning clutch
(b) A bad solenoid
(c) High resistance in the motor feed circuit
(d) High resistance in the motor control circuit

50. When the starting motor will not crank the engine or cranks it too slowly, the first thing to check is the:
(a) Battery (b) Wiring and cables
(c) Solenoid (d) Starting motor

ANSWERS

1. (b)	**2.** (c)	**3.** (a)	**4.** (d)	**5.** (a)	**6.** (a)	**7.** (a)	**8.** (c)	**9.** (a)	**10.** (b)
11. (c)	**12.** (d)	**13.** (b)	**14.** (b)	**15.** (b)	**16.** (b)	**17.** (a)	**18.** (a)	**19.** (d)	**20.** (b)
21. (c)	**22.** (d)	**23.** (d)	**24.** (d)	**25.** (d)	**26.** (b)	**27.** (c)	**28.** (c)	**29.** (b)	**30.** (b)
31. (c)	**32.** (b)	**33.** (d)	**34.** (a)	**35.** (c)	**36.** (a)	**37.** (c)	**38.** (b)	**39.** (d)	**40.** (b)
41. (a)	**42.** (d)	**43.** (c)	**44.** (d)	**45.** (a)	**46.** (c)	**47.** (d)	**48.** (a)	**49.** (a)	**50.** (a)

EXPLANATORY NOTES

1. The solenoid switch is a strong electromagnetic switch. It is used to operate the over-running clutch drive pinion to engage with the flywheel ring gear. It also acts as a relay to close the contacts between the battery and the starting motor.

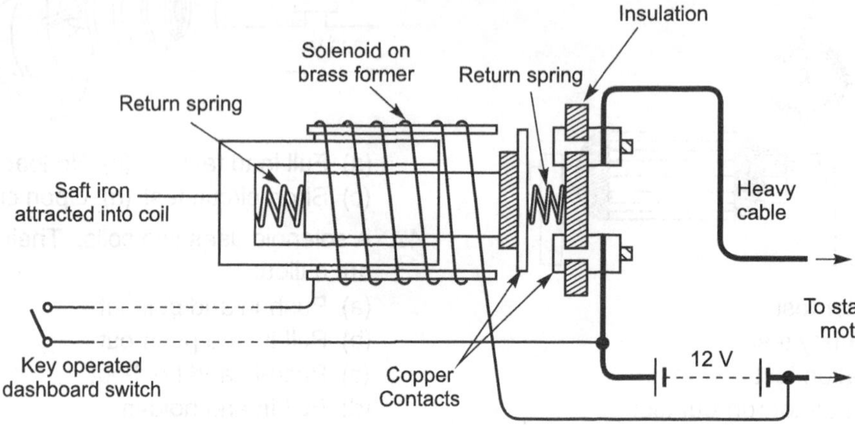

Solenoid switch

2. **Axial or sliding armature drive:** This type of drive allows its armature to slide inorder to enable its pinion to come in mesh with the flywheel ring gear. Bendix drive is the most commonly used mechanism.
3. To protect the battery and the accessories against high voltage, the alternator voltage must be controlled. This is done by using a voltage regulator which varies the current flow to the rotating field (rotor). The regulator work is done electronically.
9. In a solenoid there are two windings, a pull-in winding and a hold-in winding. The pull-in winding is wound with thick wires (series winding) and the hold-in winding is of thin wires (shunt winding). The pull-in winding is connected to the starter switch in the solenoid. The hold-in winding is connected across the switch terminal and ground. The two windings are wound around a hollow core.

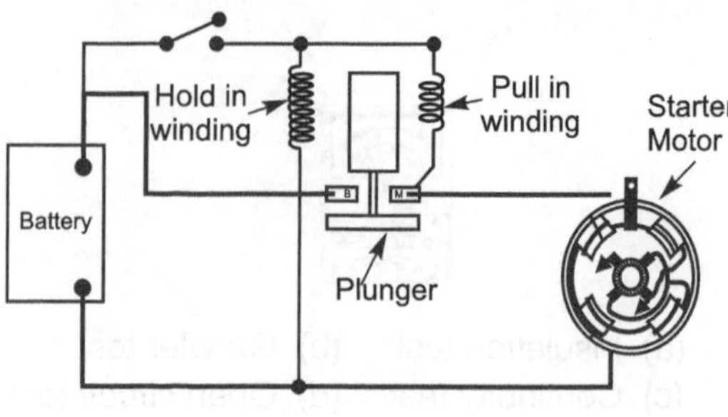

Solenoid switch

10. A growler is an electrical device used for testing insulation of a motor for shorted

coils. A growler consists of a coil of wire wrapped around an iron core and connected to a source of alternating current. When placed on the armature or stator core of a motor the growler acts as the primary of a transformer and the armature coils act as the secondary. A "feeler", a thin strip of steel (hacksaw blade) can be used as the short detector.

11. The armature is placed on a pair of V-blocks and run-out is checked using a dial gauge.

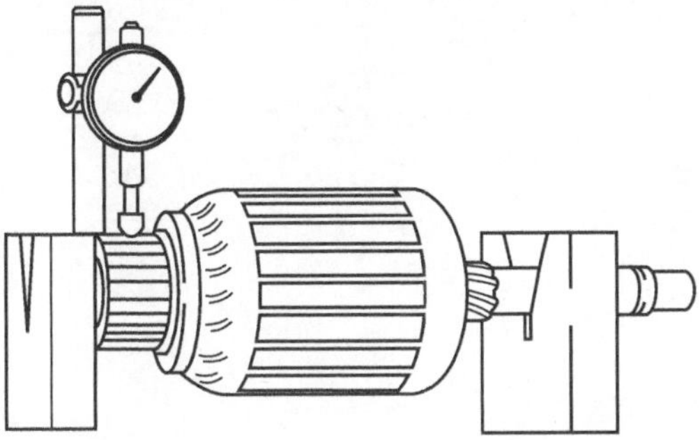

Checking run-out to commutator

19. In the starting circuit, the –ve terminal of the battery is connected to earth. The +ve terminal of the battery is connected to the solenoid switch's battery terminal. From there a wire is connected to the starter switch's input terminal. From the input terminal of the starter switch, a wire is connected to the solenoid winding's input terminal. The other end of the winding is connected to earth.

21. The figure in this question is the starting circuit of an engine.

22. The figure given in this question is of starter motor. Armature is one of its parts, which has been marked as 'X'. The motor has 2 or 4 brushes which are in contact with armature segments.

28. The figure given in this question is of overrunning clutch drive of starting system. In this the shift lever is used to slide the pinion along the armature shaft for meshing into or out of the flywheel teeth. The shift lever is operated either by a solenoid or by manual linkage.

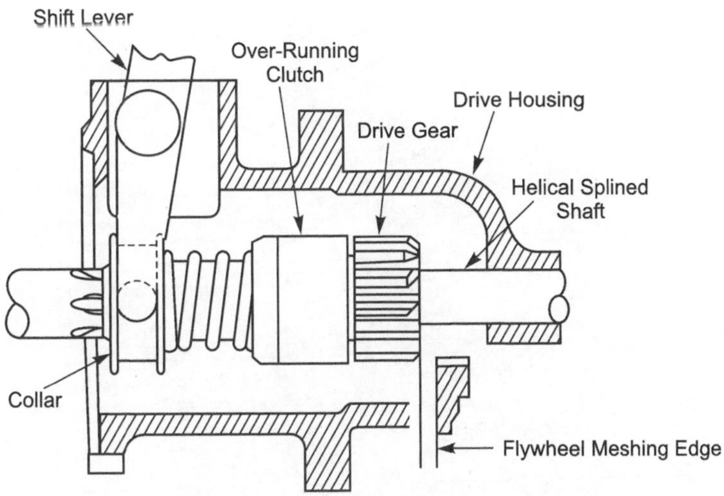

Overrunning clutch drive

32: Alternators have several advantages over direct-current generators (dynamos). They are lighter, cheaper, more rugged, and can provide useful charge at idle speed.

37: The figure given under this question is stator assembly. It is a stationary part which is held between two end covers. It consists of a laminated, cylindrical, iron core which is slotted to permit the fitting of three sets of insulated windings. In the lighter units these windings are star connected and in the heavier units delta connected.

47: Pull-in coil pulls the plunger in and the Hold-in coil holds the plunger.

49: The overrunning clutch is freewheeling in both directions.

50: A weak or faulty battery can prevent the starter from working correctly.

Module 16

Trouble Shooting

MULTIPLE CHOICE QUESTIONS

1. What is the effect of taper and ovality of bore of engine cylinder?
   (a) Loss of compression
   (b) Misfiring
   (c) Difficulty in starting
   (d) False valve timing
2. What is the cause of high fuel consumption in a diesel engine?
   (a) High oil level
   (b) Clogged air cleaner
   (c) High fuel level in tank
   (d) High compression pressure
3. What is the cause of overheating of engine?
   (a) Defective starting motor
   (b) External fuel leak
   (c) Improper injection timing
   (d) High oil pressure
4. What is the probable cause of engine low power generation?
   (a) Weak compression
   (b) Defective alternator
   (c) External leakage of oil
   (d) Clogged thermostat valve
5. Which among the following is one of the probable causes of overheating of engine?
   (a) Clogged fuel filter
   (b) Fuel leakage
   (c) External leakage of oil
   (d) Defective thermostat valve
6. What is the probable cause of high oil consumption in an engine?
   (a) Clogged air cleaner
   (b) Clogged fuel filter
   (c) Worn out piston rings
   (d) Low oil level
7. Which among the following is one of the probable causes of overheating of engine?
   (a) High air draft
   (b) High circulation of water
   (c) Blocked radiator core
   (d) High water level in radiator
8. Which among the following is one of the probable causes of overheating of engine?
   (a) High water level
   (b) High oil level
   (c) Faulty water pump
   (d) Faulty injector
9. What is the remedy if an engine overheats?
   (a) Top up oil level
   (b) Top up fuel level
   (c) Top up coolant level
   (d) Top up electrolyte level
10. What is the formula for BHP?
    (a) 2π NT / 4500 (b) 4π NT / 4500
    (c) 6π NT / 4500 (d) 8π NT / 4500
11. The device which is used to measure the brake horsepower of an engine is called:
    (a) Indicator (b) Torque wrench
    (c) Dynamometer (d) Energy meter

12. The power of an agent is said to be 1 hp, if it is working at the rate of:
(a) 33,000 ft lb / min
(b) 33,000 ft lb / sec
(c) 550 ft lb / min
(d) 550 ft lb / hour

13. The instrument which is used to obtain p – v diagram of an engine is called:
(a) Dial gauge (b) Indicator
(c) Snap gauge (d) Feeler gauge

14. What is the formula for compression ratio in an engine cylinder?
(a) (VS + VC) ÷ VC (b) (VS + VC) ÷ VS
(c) VC ÷ (VS + VC) (d) VS ÷ (VS + VC)

15. What is the formula for mechanical efficiency of an engine?
(a) BHP / IHP × 100
(b) IHP / BHP × 100
(c) (BHP – IHP) / BHP × 100
(d) (IHP – BHP) / IHP × 100

16. What is the probable cause of engine running erratically?
(a) Weak compression
(b) High compression
(c) Air lock in fuel system
(d) Poor oil circulation

17. What is the effect of defective fuel injector?
(a) Engine does not start
(b) Engine overheats
(c) High fuel pressure
(d) High oil pressure

18. What is the effect of defective oil pump?
(a) Engine is seized
(b) High oil pressure
(c) Low oil pressure
(d) Detonation occurs

19. What is to be checked if an engine overheats?
(a) Hydraulic fluid level
(b) Fuel level
(c) Coolant level
(d) Electrolyte level

20. What is to be checked for low power generation of an engine?
(a) Water level
(b) Clogged oil filter
(c) Battery condition
(d) Clogged air cleaner

21. What is the probable cause if an engine fails to start?
(a) Defective oil pump
(b) Defective thermostat
(c) Defective water pump
(d) Defective starter motor

22. What is the effect if the battery is discharged?
(a) Engine does not start
(b) Abnormal noise in the engine
(c) Engine backfires
(d) Air lock in the fuel system

ANSWERS

1. (a)	**2.** (b)	**3.** (c)	**4.** (a)	**5.** (d)	**6.** (c)	**7.** (c)	**8.** (c)	**9.** (c)	**10.** (a)
11. (c)	**12.** (a)	**13.** (b)	**14.** (a)	**15.** (a)	**16.** (c)	**17.** (a)	**18.** (a)	**19.** (c)	**20.** (d)
21. (d)	**22.** (a)								

EXPLANATORY NOTES

General Remarks: The table given below lists important causes and effects of engine problems.

Cause	Effect							
	Engine not starting	**Hard to start engine**	**Runs rough at lower RPM**	**Lack of power**	**Diesel knock/ pinking**	**Black smoke**	**White smoke**	**Blue smoke**
Low compression	X	X					X	
Low fuel pressure	X	X	X	X		X	X	
Low cranking speed – flat battery	X	X						
Glow plugs or relay faulty	X	X					X	
Insufficient fuel supply	X	X	X	X			X	
Fuel quality and contamination	X	X	X	X	X		X	
Air – vacuum in fuel supply	X	X	X	X				
Blocked fuel supply – filters	X	X	X	X	X			
Faulty diesel injector (s)	X	X	X	X	X	X	X	
Faulty high pressure pump	X	X	X	X		X		
Faulty pressure regulator – sensor	X	X	X	X		X		
Faulty low pressure supply pump	X	X	X	X				
Air intake restriction				X	X		X	
Turbo problems – waste gate and variable vanes				X	X	X	X	X
EGR problems – now including water cooled EGR valves leaking water into the engine	X	X	X	X		X	X	

Injector blow-by seat leaking			X	X	X	X		
Cam – crank sensor	X	X	X		X	X	X	
Injector wiring harness			X		X	X	X	
Internal engine problems	X	X	X	X				X

10: In this formula, BHP = Brake horsepower; N = Engine speed in rpm; T = Brake torque

11: Prony Brake is one of the simplest dynamometers for measuring power output (brake power). It attempts to stop the engine using a brake on the flywheel and measure the weight which an arm attached to the brake will support, as it tries to rotate with the flywheel.

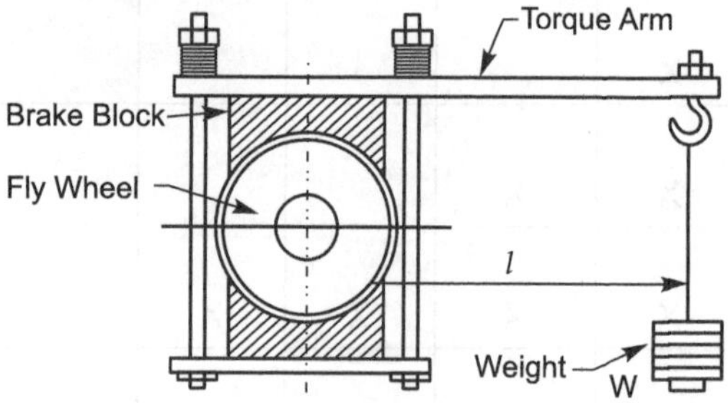

Prony brake

13: Indicated power is determined with the help of a p – V diagram taken with the help of an engine indicator. The indicated horsepower may be computed by measuring the area of the diagram, either with a planimeter or by ordinate method, and dividing by the stroke in order to obtain the mean effective pressure (m.e.p.).

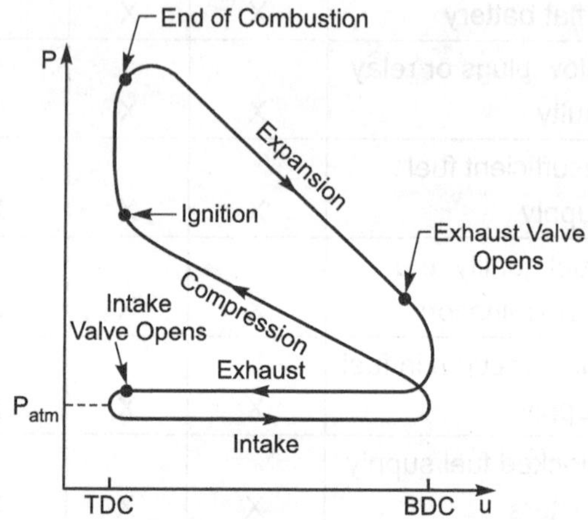

P – V diagram

14: In this formula, VS = Swept volume and VC = Clearance volume. The swept volume is the cylinder volume between T.D.C. (Top Dead Centre) and B.D.C. (Bottom Dead Centre). The clearance volume is the space above the piston when it is at T.D.C.

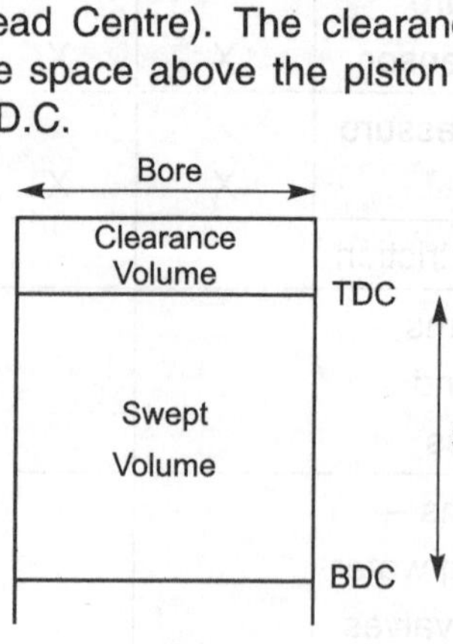

Swept volume and clearance volume

15: IHP is Indicated Horsepower; BHP is Brake Horsepower

August 2018 (Semester-1)

All India Trade Test

Mechanic Diesel Theory Paper

(Common for Mechanic (Diesel), Mechanic Tractor/Mechanic Auto Electrician and Electronics/Mechanic Motor Cycle/Pump Operator-cum-Mechanic

Time: 3 Hrs. **Part-A (Theory)** *Total Marks : 150*

Note: Attempt all questions. All questions carry equal marks.

Choose the correct answer–

Q.1. According to Ohm's law–
(a) The voltage is constant to current
(b) The voltage is proportional to the current
(c) The voltage is inversely proportional to the current
(d) The voltage is proportional to the resistance

Q.2. In a parallel circuit across each resistance–
(a) Current divides and voltage remains the same
(b) Voltage divides and the current remains the same
(c) Both (a) and (b)
(d) None of these

Q.3. Which one of the following is a semi-conductor?
(a) Gold (b) Glass
(c) Silicon (d) Plastic

Q.4. Which of the following instruments is used for checking flatness and squareness of a surface?
(a) Try square
(b) Vernier height gauge
(c) Slip gauge
(d) Bevel gauge

Q.5. The minimum measurement that can be read with the help of a steel rule is–
(a) 0.01 mm (b) 0.02 mm
(c) 0.05 mm (d) 0.50 mm

Q.6. Multimeter is used to measure–
(a) Voltage
(b) Current
(c) Resistance
(d) All of these

Q.7. In ______ valve a wedge is used to control the flow.
(a) Globe (b) Butterfly
(c) Diaphragm (d) Gate

Q.8. SCR is known as–
(a) Silicon controlled rectifier
(b) Thyristor
(c) Zener diode
(d) Both (a) and (b)

Q.9. In a metric micrometer, a complete revolution of thimble advances–
(a) 0.01 mm (b) 0.25 mm
(c) 0.50 mm (d) 1.00 mm

Q.10. Ratchet stop in the micrometer helps to–
(a) Control the pressure
(b) Lock the spindle
(c) Adjust zero error
(d) Hold the workpiece

Q.11. A feeler gauge is used for–
(a) Adjusting the parts for play
(b) Checking the gap between mating parts
(c) Checking radii of job
(d) Checking accuracy of hole

Q.12. The point angle of centre punch is–
(a) 30° (b) 60°
(c) 90° (d) 120°

Q.13. The purpose of normalizing is to–
(a) Remove induced stresses
(b) Improve machinability
(c) Soften the steel
(d) Increase toughness

Q.14. Eye foundation bolt is also known as–
(a) Hook bolt
(b) Rag bolt
(c) Cotter bolt
(d) Lewis bolt

Q.15. Best insulation among the following is–
(a) Copper (b) Aluminium
(c) Iron (d) Mica

Q.16. For general purpose, a bench vice is fixed at a height of–
(a) 80 cm (b) 90 cm
(c) 106 cm (d) 125 cm

Q.17. Which one of the following tools is perfect for quickly loosening and tightening nuts and bolts?
(a) Air impact wrench
(b) Air Ratchet
(c) Ring spanner
(d) Socket Spanner

Q.18. Which one of the following is used when a precise amount of torque is to be imparted to the fastener?
(a) Air ratchet
(b) Ring spanner
(c) Torque wrench
(d) Monkey wrench

Q.19. Which one among the following scrapers is used for scraping on bearing bush?
(a) Flat scraper
(b) Triangular scraper
(c) Half round scraper
(d) Round scraper

Q.20. Two way pliers is necessary to operate a–
(a) Castle nut (b) Wring nut
(c) Locking nut (d) Circlip

Q.21. File card is used for–
(a) Cleaning the workpiece
(b) Cleaning the file teeth
(c) Renewing the file teeth
(d) Cleaning the chips

Q.22. A transistor has–
(a) One PN junction
(b) Two PN junction
(c) Three PN junction
(d) Four PN junction

Q.23. A cutting tool used to finish and enlarge a hole is known as–
(a) Drill (b) Tap
(d) Die (c) Reamer

Q.24. Bent snip is used for–
(a) Removing the burr
(b) Making hole on the sheet
(c) Bending the sheet
(d) Cutting the sheet along curved lines

Q.25. A cutting tool used to cut outside threads is called–
(a) Drill (b) Reamer
(c) Die (d) Tap

Q.26. Which one of the following is used to rectify damaged or rusted threads?
(a) Die nut (b) Circular split die
(c) Two-piece die (d) Die plate

Q.27. What is used for removing a broken tap?
(a) Tap disposer (b) Tap wrench
(c) Tap extractor (d) None of these

Q.28. The grooves provided on the entire length of the body of the twist drill are called–
(a) Lips (b) Flutes
(c) Margins (d) Webs

Q.29. A wedge like tool is used for removing the taper shank tools from the nose of the machine spindle. What is this called?
(a) Drill chuck (b) Drill key
(c) Drill socket (d) Drill drift

Q.30. Hacksaw blade teeth get dull due to–
(a) High speed and pressure
(b) Pressure not released during return stroke
(c) Coolant not used
(d) Anyone of the above

Q.31. The point angle of twist drill depends upon the–
(a) Cutting speed
(b) Type of drilling machine
(c) Size of the drill
(d) Material to be drilled

Q.32. The point angle of twist drill for general purpose is–
(a) 135° (b) 118°
(c) 90° (d) 60°

Q.33. Drill chucks are held on the machine spindle by means of–
(a) Arbor (b) Drift
(c) Draw-in bar (d) Chuck nut

Q.34. Which one of the following machine reamers is used to correct the misalignment between the reamer axis and the work axis?
(a) Floating blade reamer
(b) Shell reamer
(c) Machine jig reamer
(d) Chucking reamer

Q.35. Which one of the following drilling machines is used for drilling holes where electricity is not available?
(a) Bench drilling machine
(b) Pillar drilling machine
(c) Radial drilling machine
(d) Ratchet drilling machine

Q.36. A nut is to be made for a screw of M10 × 1.0. What should be the size of drilled hole?
(a) 8.5 mm (b) 9.0 mm
(c) 9.5 mm (d) 10.0 mm

Q.37. Which among the following tools is used for flattening the metal around punched hole?
(a) Ball peen hammer
(b) Riveting hammer
(c) Setting hammer
(d) Sledge hammer

Q.38. Besides a CO_2 fire extinguisher, which type of fire extinguisher can you use on electrical fires?
(a) Foam
(b) CTC extinguisher
(c) Dry chemical extinguisher
(d) Soda acid extinguisher

Q.39. Best spanner for automobile work is–
(a) Open ended type
(b) Combination type
(c) Ring type
(d) Socket type

Q.40. For tightening the cylinder head bolts, we should prefer to use–
(a) Open ended spanner
(b) Torque wrench
(c) Ring spanner
(d) Adjustable spanner

Q.41. Taper shank twist drill are provided with Morse taper ranging from–
(a) MT1-MT4 (b) MT1-MT5
(c) MT1-MT10 (d) None of these

Q.42. File removes the metal during–
(a) Forward stroke (b) Return stroke
(c) Both (a) and (b) (d) None of these

Q.43. The welding machine, which is used to convert AC welding supply to DC welding supply is–
(a) Motor generator set
(b) Engine generator set
(c) Rectifier set
(d) Welding transformer

Q.44. Phillips screw driver is used when–
(a) More finished work is required
(b) Less finished work is required
(c) Rough work is required
(d) Any type of work is required

Q.45. Screws and bolts which are broken below the surface, are removed by–
(a) Hammer and chisel
(b) Screw driver
(c) Pliers
(d) Screw driver extractors

Q.46. The benefit of 5S system is–
(a) Increase production rate
(b) Increase cleanliness
(c) Increase productivity
(d) None of these

Q.47. Which one of the following logic valve is known as shuttle valve?
(a) OR gate (b) AND gate
(c) NOR gate (d) NAND gate

Q.48. Lapping is a process for finishing______
(a) External surface only
(b) Internal surface only
(c) External and internal surfaces
(d) Flat surface only

Q.49. Solder is an alloy consisting of–
(a) Tin, antimony, copper
(b) Tin and copper
(c) Tin and lead
(d) Lead and zinc

Q.50. The composition of silver solder is–
(a) Silver, copper, zinc
(b) Silver, tin, nickel
(c) Silver, lead, zinc
(d) Silver, copper, aluminium

Key to Aug. 2018 (Semester-1) Mechanic Diesel Theory Paper

1. (b)	**2.** (a)	**3.** (c)	**4.** (a)	**29.** (d)	**30.** (d)	**31.** (d)	**32.** (b)
5. (d)	**6.** (d)	**7.** (d)	**8.** (d)	**33.** (a)	**34.** (a)	**35.** (d)	**36.** (b)
9. (c)	**10.** (a)	**11.** (b)	**12.** (c)	**37.** (a)	**38.** (c)	**39.** (d)	**40.** (b)
13. (a)	**14.** (a)	**15.** (d)	**16.** (c)	**41.** (b)	**42.** (a)	**43.** (c)	**44.** (a)
17. (b)	**18.** (c)	**19.** (c)	**20.** (d)	**45.** (d)	**46.** (c)	**47.** (a)	**48.** (c)
21. (b)	**22.** (b)	**23.** (d)	**24.** (d)	**49.** (c)	**50.** (a)		
25. (c)	**26.** (a)	**27.** (c)	**28.** (b)				

August 2018 (Semester-2)

All India Trade Test

Mechanic Diesel Theory Paper

Time: 3 hrs. *Total Marks : 150*

Note: Attempt all questions. All questions carry equal marks.

Choose the correct answer–

Q.1. Engine develops less power due to–
(a) Defective ignition timing
(b) Excessive rich mixture
(c) Defective lubrication system
(d) Too tight cylinder head

Q.2. The starter motor is driven by–
(a) Chain drive (b) Gear drive
(c) Flat belt drive (d) V-belt drive

Q.3. A glow plug is–
(a) A compression device
(b) A decompression device
(c) Fitted in the inlet fold
(d) Screwed in the combustion chamber

Q.4. A 4-stroke engine produces one power stroke in–
(a) 2 revolutions of crankshaft
(b) 4 revolutions of crankshaft
(c) 6 revolutions of crankshaft
(d) 8 revolutions of crankshaft

Q.5. In an engine stroke length is–
(a) Half the throw of crank
(b) Equal to throw of crank
(c) Double the throw of crank
(d) Three times the throw of crank

Q.6. The high fuel consumption of a petrol engine may be due to–
(a) Leakage of fuel from carburettor
(b) Defects in lubrication system
(c) Air leaks in intake manifold
(d) Incorrect idle speed

Q.7. The stroke of an engine is–
(a) Inside diameter of the cylinder
(b) Distance between TDC and BDC
(c) Volume of the cylinder
(d) Length of the connecting rod

Q.8. Air-fuel mixture in a petrol engine is burnt in–
(a) intake manifold
(b) Exhaust manifold
(o) Cylinder
(d) None of these

Q.9. Which one of the following types of engines has the longest crankshaft?
(a) Opposed engine(b) Radial engine
(c) In-line engine (d) 'V' engine

Q.10. In a diesel engine, fuel and air mix together in the–
(a) Carburettor
(b) injector
(c) Combustion chamber
(d) Inlet ports

Q.11. In diesel engine, the fuel is ignited by–
(a) A glow plug
(b) A spark plug
(c) An injector
(d) Virtue of temperature of compressed air

Q.12. A hole is drilled between crankshaft main journal and crank pin for–
(a) Balancing of crankshaft
(b) Reducing crankshaft vibration
(c) Reducing crankshaft weight
(d) Lubricating connecting rod bearing

Q.13. Scavenging is done to maintain the–
(a) Power output
(b) Mechanical efficiency
(c) FHP
(d) Oil pressure

Q.14. In a turbocharged engine, the air pressure is increased by turbine-compressor driven by–
(a) Crankshaft (b) Fan pulley
(c) Starter motor (d) Exhaust gases

Q.15. In a SI engine, the spark plug fires just–
(a) Before the piston reaches the bottom
(b) After the piston reaches the top
(c) Before the piston reaches the top
(d) After the piston reaches the bottom

Q.16. Crankshaft bearings are made in two halves because they are–
(a) Easy to manufacture
(b) Easy to replace
(c) Cheap to manufacture
(d) Wear less

Q.17. A wrist pin connects–
(a) Piston to the connecting rod
(b) Connecting rod to the crankshaft
(c) Crankshaft to the camshaft
(d) Camshaft to the push rod

Q.18. The combustion chamber is the area between–
(a) TDC and BDC
(b) Top of piston and TDC
(c) Top of piston and BDC
(d) None of these

Q.19. All petrol engines have a .... valve to control the volume of intake air.
(a) Choke (b) Throttle
(c) Float (d) Venturi

Q.20. Which engine has maximum fuel efficiency?
(a) V engine
(b) In-line engine
(c) Radial engine
(d) Opposed engine

Q.21. In diesel cycle, combustion takes place at–
(a) Constant pressure
(b) Constant volume
(c) Constant temperature
(d) Constant temperature and pressure

Q.22. The spark occurs when–
(a) Points close
(b) Points open
(c) Ignition switch is on
(d) None of these

Q.23. The contact breakage points are opened by the cam and closed by the–
(a) Spring tension (b) Centrifugal force
(d) Magnetic force (c) None ot these

Q.24. Tachometer in a diesel engine measures–
(a) Speed (b) Distance
(c) Engine rpm
(d) Fuel consumption

Q.25. Which one of the following in a diesel engine fuel system is located just before the injector?
(a) Fuel tank (b) Fuel lifter
(c) Fuel filter (d) Lift pump

Q.26. To which one of the following parts of the diesel engine injector delivers fuel?
(a) Fuel tank (b) Fuel filter
(c) Engine cylinder (d) Governor

Q.27. The fuel injection timing in a distributor type pump is controlled by–
(a) Changing plunger stroke
(b) Changing speed of rotor
(c) Rotating the cam ring
(d) Changing the number of cams on the ring

Q.28. The purpose of the injector provided on a diesel engine is to inject–
(a) Lubricant in the cylinder
(b) Lubricant in the bearings
(c) Fuel in the cylinder
(d) Coolant in the cylinder jacket

Q.29. The gap of a spark plug is adjusted by
(a) Filing the centre electrode
(b) Filing the earth electrode
(c) Bending the centre electrode
(d) Bending the earth electrode

Q.30. The purpose of the fan in engine cooling system is to–
(a) Increase flow of coolant
(b) Draw air through the radiator
(c) Cool the engine by blowing air over it
(d) Provide drive to the water pump

Q.31. The purpose of thermostat is to keep the engine–
(a) Cool
(b) Warm
(c) Hot
(d) At desired temperature

Q.32. Which one of the following gives the correct location of thermostat in the cooling system?
(a) Between header hose and radiator
(b) Between radiator and bottom hose
(c) Between hose and engine water jacket
(d) Between engine water jacket and header hose

Q.33. Which part indicates the oil level in the engine sump?
(a) Feeler gauge
(b) Plastic gauge
(c) Oil pressure gauge
(d) Dip stick

Q.34. The purpose of piston pin is to–
(a) Prevent the valve from rotating
(b) Link the connecting rod to the crankshaft
(c) Ensure the piston ring to the piston
(d) Connect the piston to the connecting rod

Q.35. The two types of lubricating system used in engines are–
(a) High and low pressure
(b) Pressure and vacuum
(c) Pump and gravity
(d) Splash and pressure

Q.36. The radiator core is made of–
(a) Plastic (b) Cast iron
(c) Steel (d) Brass

Q.37. In a diesel engine a clogged nozzle could result in–
(a) Excessive fuel consumption
(b) Over-rich mixture
(c) Excessive smoke
(d) Any one of these

Q.38. Super charging in a diesel engine is also called–
(a) Scavenging (b) Boosting
(c) Cleaning (d) Silencing

Q.39. Which type of filters is used in large diesel engines?
(a) Dry filters
(b) Impingement filters
(c) Oil-bath filters
(d) Any one of these

Q.40. The exhaust manifold in small diesel engines is made of–
(a) Cast iron (b) Brass
(c) Aluminium (d) Copper

Q.41. The volume of muffler is usually–
(a) Two to four times the engine displacement volume
(b) Six to eight times the engine displacement volume
(c) Ten to twelve times the engine displacement volume
(d) Fourteen to sixteen times the engine displacement volume

Q.42. Deposit of carbon in the exhaust system–
(a) Will increase back pressure
(b) Will reduce back pressure
(c) Will have no effect on back pressure
(d) Will result in fluctuating pressure

Q.43. A pressure cap contains a–
(a) Pressure valve
(b) Thermostat valve
(c) Blow-off valve
(d) Pressure and vacuum valve

Q.44. Which one of the following type of the thermostats is most commonly used in modern engines?
(a) Wax-pellet type (b) Bi-metal type
(c) Bellow type (d) Alcohol type

Q.45. If the thermostat is stuck closed, the engine will–
(a) Warm up slowly (b) Oveheat
(c) Fail to start (d) Idle

Q.46. The diesel engine usually emits excessive smoke during–
(a) Starting (b) Idling
(c) Accelerating (d) Decelerating

Q.47. Which one of the following fuel injection systems was originally used in large stationary and marine engines?
(a) Air injection
(b) Airless injection
(c) Mechanical injection
(d) Solid injection

Q.48. A smoky blue exhaust is due to–
(a) Lean mixture
(b) Excessive fuel consumption
(c) Burning of oil in the combustion chamber
(d) Incorrect valve adjustment

Q.49. Series winding of solenoid is generally made of–
(a) Thin wire (b) Thick wire
(c) Flex wire (d) PVC wire

Q.50. Engine does not start because of–
(a) Absence of fuel filters
(b) Punctured float
(c) Less water in radiator
(d) Low viscosity of engine oil

Key to Aug. 2018 (Semester-2) Mechanic Diesel Theory Paper

1. (a)	**2.** (b)	**3.** (d)	**4.** (a)	**29.** (d)	**30.** (b)	**31.** (d)	**32.** (d)
5. (c)	**6.** (a)	**7.** (b)	**8.** (c)	**33.** (d)	**34.** (d)	**35.** (d)	**36.** (d)
9. (c)	**10.** (c)	**11.** (d)	**12.** (d)	**37.** (d)	**38.** (b)	**39.** (b)	**40.** (a)
13. (a)	**14.** (d)	**15.** (c)	**16.** (b)	**41.** (b)	**42.** (a)	**43.** (d)	**44.** (a)
17. (a)	**18.** (b)	**19.** (b)	**20.** (c)	**45.** (b)	**46.** (c)	**47.** (a)	**48.** (c)
21. (a)	**22.** (c)	**23.** (a)	**24.** (c)	**49.** (c)	**50.** (b)		
25. (c)	**26.** (c)	**27.** (c)	**28.** (c)				

August 2019 (Annual)

All India Trade Test

Mechanic Diesel Theory Paper

Total Time : 2 hrs. *Total Marks : 100*

***Note:* Attempt all questions. All questions carry equal marks.**

Choose the correct answer.

Q.1. What material is used for making a file?

(a) High Carbon Steel

(b) Stainless Steel

(c) Mild Steel

(d) Sponge Iron

Q.2. Which of the following part of engine stores the energy during power stroke –

(a) Vibration damper

(b) Crankshaft

(c) Flywheel

(d) Camshaft

Q.3. Tempering is the process in which –

(a) Toughness of the metal is increased by reducing its brittleness

(b) Cuttng edge is hardened

(c) Cutting edge is sharpened

(d) Cutting edge is checked

Q.4. Tap is used to make ...... threads.

(a) Potential (b) External

(c) Central (d) Inside

Q.5. There are ...... types of nozles used in the engine.

(a) 4 (b) 5

(c) 6 (d) 7

Q.6. Which one is not a type of scrapper?

(a) Half-round (b) Triangular

(c) Rectangular (d) Flat

Q.7. Name part of file where a handle is fastened for the smooth working –

(a) Shoulder (b) Tang

(c) Heel (d) Toe

Q.8. Connecting rod is made of –

(a) Spring steel (b) Forged steel

(c) Stainless steel (d) Mild steel

Q.9. Oil pump is fitted in an engine to suck oil from sump, and send it directly to –

(a) Oil gallery (b) Oil filters

(c) Oil gauge (d) Oil strainer

Q.10. Sheet/wire gauges are used to measure –

(a) Strength (b) Malleability

(c) Tenacity (d) Thickness

Q.11. The engine valves are closed by –

(a) Crankshaft (b) Camshaft

(c) Timing device (d) Valve springs

Q.12. If thermostat valve remains stuck in closed position which of following will happen –

(a) Rough Idling speed

(b) Engine will warm up slowly

(c) Starting will be difficult

(d) Engine will overheat

Q.13. In which type of engine crankshaft length is maximum for the same numbers of cylinder?

(a) V-Engine

(b) Inline Engine

(c) Radial Engine

(d) Opposed cylinder Engine

Q.14. During overlapping period inlet and exhaust valves are –

(a) Open

(b) Closed

(c) Inlet opening Exhaust closing

(d) Exhaust opening Inlet closing

Q.15. Which fire extinguisher (FE) is suitable for a live electric fire?

(a) Soda acid FE

(b) Water filled FE

(c) Foam FE

(d) C.T.C. FE

Q.16. In positive plate ...... is filled while negative plates have ....

(a) Iron, Copper

(b) Lead peroxide, lead antimony

(c) Brass, Aluminium

(d) Hydrochloride, Copper sulphate

Q.17. What are the losses in transformer?

(a) Cooling loss

(b) Winding loss

(c) Iron loss & Copper loss

(d) Circuit loss

Q.18. Reverse voltage at which junction breaks is called –

(a) High voltage

(b) Breakdown voltage

(c) Low voltage

(d) None of these

Q.19. In arc welding which machine converts A.C. supply into D.C. supply –

(a) Transformer (b) Blowpipe

(c) Rectifier (d) None of these

Q.20. With the use of ...... 20% to 30% more power from engine is achieved.

(a) Coolant (b) Supercharger

(c) CNG (d) High grade oil

Q.21. The diesel engines are called –

(a) C.I. engines

(b) S.I. engines

(c) Gasoline engines

(d) X.Y. engines

Q.22. Glow plugs are fitted in –

(a) Direct injection system

(b) Indirect injection system

(c) Both (a) & (b)

(d) None of these

Q.23. How many degrees before TDC in compression stroke in diesel engine fuel injection take place –

(a) 33 degrees (b) 23 degrees

(c) 15 degrees (d) 10 degrees

Q.24. Diesel engine emit more ...... and ...... than petrol engines.

(a) Particulates & Sulphur oxides

(b) Hydrocarbon & CO

(c) Carbon & Water

(d) Oxygen & Carbon

Q.25. What is the cetane number of diesel?

(a) 0 to 15 (b) 20 to 35

(c) 35 to 65 (d) 65 to 75

Q.26. The power of an engine available at fly wheel is called ....

(a) HP (b) BHP

(c) IHP (d) None of these

Q.27. Incomplete combustion of fuel may be due to ....

(a) Wrong injection timing

(b) Defective injector

(c) Defective F.I.P.

(d) All of these

Q.28. There are ...... types of nozzles used in the engine.

(a) 4 (b) 5

(c) 6 (d) 7

Q.29. Turbo charger mounted on the engine helps to ....
(a) Burn less fuel
(b) Burn more fuel
(c) Maintain engine power
(d) Decrease engine power

Q.30. Due to detonation, ...
(a) No sound is heard in the engine
(b) A knocking sound is heard in the engine
(c) Normal combustion occurs in the engine
(d) None of these

Q.31. Main function of a lubricant is to ....
(a) Maximise the contact between the moving surfaces by the layers of lubricant
(b) Minimise the friction between the two moving surfaces in contact
(c) Release heat from the moving parts due to friction
(d) Provide resistance between the moving parts

Q.32. Oil pump may be of ...... types.
(a) 2 (b) 3
(c) 4 (d) 5

Q.33. Pressure cap has ....
(a) Pressure valve only
(b) Vacuum valve only
(c) Both (a) & (b)
(d) No valve

Q.34. Combustion of fuel inside a cylinder develops a temperature of about ....
(a) 22°C (b) 220°C
(c) 2200°C (d) 22000°C

Q.35. Chain drive is also called as ....
(a) Belt drive (b) Rope drive
(c) Sprocket drive (d) None of these

Q.36. Cylinder head has passages for ...... circulation.
(a) Oil (b) Water
(c) Oil & water (d) None of these

Q.37. Valve timing refers to ....
(a) Opening and closing of the valve in an IC engine in relation to the movement of the piston and flywheel
(b) Opening and closing of the valve in an IC engine in relation to the movement of the piston
(c) Opening and closing of the valve in an IC engine in relation to the movement of the flywheel
(d) None of these

Q.38. Crankshaft transmits ...... to the ....
(a) Reciprocating motion, rotary motion
(b) Torque, flywheel
(c) Centrifugal force, connecting rod
(d) Centrifugal force, piston

Q.39. A crankshaft withstands ....
(a) Centrifugal force only
(b) Impact force by the piston only
(c) Impact force by the piston and the connecting rod
(d) Both (a) & (c)

Q.40. The ...... end of connecting rod is fitted with a phosphor bronze bush.
(a) Small
(b) Big
(c) Either (a) or (b)
(d) Neither (a) nor (b)

Q.41. Blow of gases are largely ...... gases.
(a) CO (b) HC
(c) NO_X (d) PM

Q.42. DC starter motor may be of ...... kind.
(a) Series (b) Shunt
(c) Compund (d) All of these

Q.43. Marine engines are ....

(a) Petrol engine only

(b) Diesel engine only

(c) Both (a) or (b)

(d) None of these

Q.44. High pressure is exerted on the piston in the .... stroke and resistance of piston is in the .... strokes.

(a) Compression, power

(b) Power, compression

(c) Power, power

(d) Compression, compression

Q.45. A device for holding any speed steady between idling and maximum speed is called ....

(a) Carburettor (b) Flywheel

(c) Governor (d) Fuel pump

Q.46. Sulphur present in fuel ...... corrosion of engine.

(a) Decreases (b) Increases

(c) Maintains (d) None of these

Q.47. Oil pump may be of ...... types.

(a) Rotor (b) Gear

(c) Vane (d) All of these

Q.48. The additional area for oxidation or combustion to occur in automatic emission is called ....

(a) Electronic muffler

(b) Back pressure muffler

(c) Back pressure

(d) Catalytic converter

Q.49. Oil pump is located in ....

(a) Radiator (b) Crankcase

(c) Governor (d) None of these

Q.50. Indirect cooling of engines refers to ....

(a) Air cooling only

(b) Water cooling only

(c) Both (a) or (b)

(d) None of these

Key to August 2019 (Annual) Mech. Diesel Theory Paper

1. (a)	**2.** (c)	**3.** (a)	**4.** (d)	**5.** (c)	**26.** (b)	**27.** (d)	**28.** (c)	**29.** (b)	**30.** (b)
6. (c)	**7.** (b)	**8.** (b)	**9.** (b)	**10.** (d)	**31.** (b)	**32.** (c)	**33.** (c)	**34.** (c)	**35.** (c)
11. (d)	**12.** (d)	**13.** (b)	**14.** (a)	**15.** (d)	**36.** (c)	**37.** (a)	**38.** (b)	**39.** (d)	**40.** (a)
16. (b)	**17.** (c)	**18.** (a)	**19.** (c)	**20.** (b)	**41.** (b)	**42.** (d)	**43.** (c)	**44.** (b)	**45.** (c)
21. (a)	**22.** (a)	**23.** (b)	**24.** (a)	**25.** (c)	**46.** (b)	**47.** (d)	**48.** (d)	**49.** (b)	**50.** (b)